IB STUDY GUIDES

Mathematical Studies

FOR THE IB DIPLOMA

Standard and Higher Level

Scott Genzer

OXFORD
UNIVERSITY PRESS

Great Clarendon Street, Oxford OX2 6DP

Oxford University Press is a department of the University of Oxford. It furthers the University's objective of excellence in research, scholarship, and education by publishing worldwide in

Oxford New York

Auckland Cape Town Dar es Salaam Hong Kong Karachi Kuala Lumpur Madrid Melbourne Mexico City Nairobi New Delhi Shanghai Taipei Toronto

With offices in

Argentina Austria Brazil Chile Czech Republic France Greece Guatemala Hungary Italy Japan Poland Portugal Singapore South Korea Switzerland Thailand Turkey Ukraine Vietnam

First published 2009

British Library Cataloguing in Publication Data

Data available

ISBN 978-0-19-915242-1

10 9 8 7 6 5 4 3 2 1

Acknowledgements

Cover image reproduced courtesy of Colin Anderson/Corbis

Cartoons by Andrew Painter

We have tried to trace and contact all copyright holders before publication. If notified the publishers will be pleased to rectify any errors or omissions at the earliest opportunity.

Printed in Great Britain by Bell and Bain Ltd., Glasgow

Paper used in the production of this book is a natural, recyclable product made from wood grown in sustainable forests. The manufacturing process conforms to the environmental regulations of the country of origin.

Contents

Introduction

Hello! My name is Scott and I am an IB Maths Studies SL teacher and an IB Maths Studies examiner. This is not a textbook – it is an **exam review book**. So I won't spend any time here explaining *why* things work or tell you stories about Newton and Fermat. This is a *"what do I need to know in order to do well in the Maths Studies exam"* book. It's not a formal book as you can tell. I have tried very hard to make this book student-friendly! If you like formal, stuffy textbooks, this book will not be good for you. ☺ I'm sure you won't *like* this book (who likes maths books?), but I hope you will at least be able to *use* this book in order to study for the exam.

Throughout this book I provide many hints and suggestions in order to help you earn precious marks that students often lose. Pay attention to my cast of characters:

In this book I exclusively use the Texas Instruments TI-83 / TI-84 Plus calculators as my GDC. I know that some schools and students use the Casio, but I chose to use the TI for three reasons: 1) it's by far the most popular GDC in use by IB students; 2) there isn't space to show both kinds of keystrokes and screenshots in this book; 3) Texas Instruments pays me megabucks for the exclusive rights (just kidding ☺). If you use a Casio, the screenshots should still be helpful but otherwise you'll have to figure out the keystrokes on your own. Sorry!

Lastly, remember that this is *your* exam review book. Write in it! Tear it up. Eat it. Throw it at your dog. Do what ever you like to it. With maths you have to get into it a bit. Don't treat this book like your prized copy of *The Canterbury Tales*. After the exam you should celebrate by destroying this book in a creative way. Post a video on YouTube and send me the link. I'd love to see it.

Good luck!

Dedication

This book is dedicated to . . .

the hundreds of students I have taught over the past sixteen years who have been my inspiration, laughed at my bad jokes and made me smile

Chris and Hana, who took time out of their senior year to help me write this book

Carolyn Lee and Fern Watson at OUP and my copy-editor Lyn Imeson for their infinite patience with me as a novice author

and, most importantly,

Kim

without whom none of this would have been possible. Je t'aime.

SG
September 2008

email: sgenzer@gmail.com

Facebook: search for 'Scott Genzer'

Structure of the IB Mathematical Studies SL Exam

The IB Math Studies written exam consists of two papers: Paper 1 and Paper 2. (Isn't the IB very creative? ☺)

Paper 1 has 15 short questions, each one worth 6 marks each for a total of 90 marks on the paper. All questions are of relatively equal difficulty – question 15 is not necessarily harder than question 1. You have a 5-minute reading period and then 90 minutes to do the paper. All answers are written on the exam. Calculators and formula booklets are allowed.

Paper 2 has 5 longer questions, each one worth different amounts (usually 15 to 20 marks each) and sometimes containing a Part A and a Part B. (I've never seen a Part C but theoretically it could happen!) There are still 90 total marks on the paper. Some questions are worth more than others but, on average, a student should spend around 1 minute per mark on each question. For example, if a question is worth 17 marks, s/he should spend approximately 17 minutes on the question. All questions are of relatively equal difficulty **but** the parts within each question generally go from easy to hard (i.e. part (a) is usually easier than part (f)). You have a 5-minute reading period and then 90 minutes to do the paper. All answers are written on separate answer sheets, not on the exam. Calculators and formula booklets are allowed.

Between Papers 1 and 2, the whole syllabus is generally covered. This information can really be used to your advantage. I highly recommend that you get together with your friends and/or your teacher after you take Paper 1 and "debrief" the exam. Go through this book and check off all the topics that were covered. If you see major topics that were not on Paper 1, they will most likely occur on Paper 2! I've been doing this with my students for many years and in general we are able to "predict" at least 3 out of the 5 questions that occur the next day.

Paper 1

15 short questions

6 marks each

90 marks total

5-minute reading period + 90 minutes to work

All answers written on exam

Paper 2

5 long questions

approx 15–20 marks each

90 marks total

5-minute reading period + 90 minutes to work

All answers written on separate answer sheets

Use the reading period to your advantage! Go through the entire paper and strategise what order you want to do the questions. You can't write anything during the reading period (or use your calculator!), but you can make a mental note to yourself about which questions to do first and which questions to do last.

Scoring of the IB Mathematical Studies SL Exam

Scoring on the IB Math Studies exam is fairly straightforward. Each paper is worth 90 marks. Your school will be assigned two examiners – one for each paper. Your examiners will mark your papers out of 90 marks and send those scores to the IB. After all the exams are marked, moderated and checked, the Chief Examiner assigns "grade boundaries" for each paper, and the project which you did earlier. These grade boundaries convert your raw score to an IB grade from 1 to 7. Finally, all three grades are combined together, a final set of grade boundaries are determined and final IB grades are awarded.

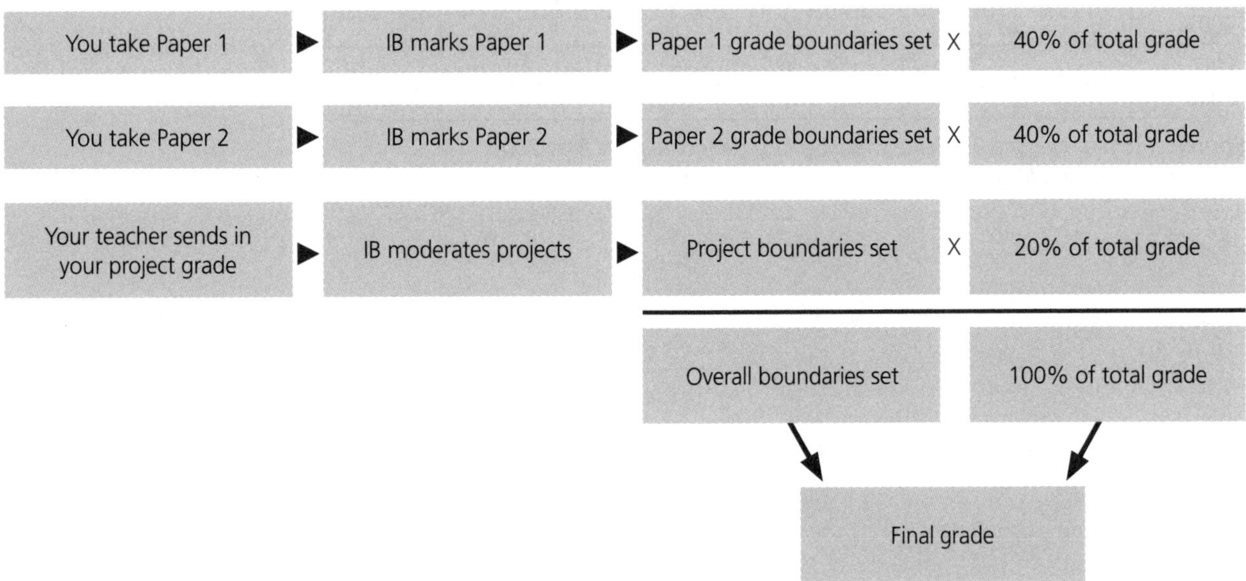

The grade boundaries vary every year so it's hard to work out what marks you need to earn a particular grade. The tables below give you some idea of what to expect based on what has been done in the past. However, be aware that the IB can set the boundaries to anything they want in a particular year!

Grade boundaries for Paper 1 – Fifteen short questions (total out of 90)							
	1	**2**	**3**	**4**	**5**	**6**	**7**
May 2007	0–12	13–25	26–32	33–45	46–57	58–70	71–90
Nov 2007	0–13	14–27	28–38	39–50	51–63	64–75	76–90
May 2008	0–11	12–22	23–33	34–46	47–59	60–72	73–90

Grade boundaries for Paper 2 – Fifteen long questions (total out of 90)							
	1	**2**	**3**	**4**	**5**	**6**	**7**
May 2007	0–11	12–22	23–29	30–40	41–52	53–63	64–90
Nov 2007	0–11	12–23	24–35	39–48	49–61	62–74	75–90
May 2008	0–9	10–19	20–28	29–40	41–53	54–65	66–90

Grade boundaries for project (out of 20)							
	1	**2**	**3**	**4**	**5**	**6**	**7**
May 2007	0–4	5–6	7–8	9–11	12–14	15–16	17–20
Nov 2007	0–4	5–6	7–8	9–11	12–14	15–16	17–20
May 2008	0–4	5–6	7–8	9–11	12–14	15–16	17–20

Overall grade boundries							
	1	**2**	**3**	**4**	**5**	**6**	**7**
May 2007	0–14%	15–27%	28–35%	36–49%	50–62%	63–75%	76–100%
Nov 2007	0–15%	16–28%	29–40%	41–55%	56–69%	70–82%	83–100%
May 2008	0–13%	14–24%	25–35%	36–49%	50–64%	65–77%	78–100%

Source: *International Baccalaureate Organisation, Subject Reports May 2007, November 2007 and May 2008*

Hints from Irene, the IB Examiner

Hint #1: Watch out for penalties

There are a few penalties that can cost you one mark each. Don't let the examiner deduct them from you. Each of the following penalties can be applied **once per paper!**

- Accuracy Penalty (AP) – this is applied when you round an answer incorrectly. It usually happens when no accuracy is specified and the student does not round to 3sf. Remember, you must *always* round to 3sf unless otherwise specified in the problem!
- Unit Penalty (UP) – this is applied when you write the answer to a problem (usually a triangle trig problem but it could be anywhere) and you forget to write the units afterwards, e.g. 3.00 instead of 3.00 cm. This is a relatively new penalty but always enforced. Don't forget the units!
- Finance Penalty (FP) – this is applied when you round an answer incorrectly in a financial math problem where the accuracy is specified. For example, the problem says to round to the nearest dollar and the student writes $15.23 instead of $15. Usually the accuracy is written **in bold letters** so read them carefully!
- Graph penalty – this isn't actually an official penalty but is a similar idea. If you draw a graph and do not *label your axes with x,y or something similar* and *indicate some kind of scale,* you will lose one mark per graph!

Hint #2: Be kind to the examiner's old eyes

It's highly likely that your examiner is at least twice your age (do the maths!), overworked (has a full-time job and is marking more than 200 exam papers) and underpaid (the IB does not pay very much for marking). Just have this image in your mind as you start your exam:

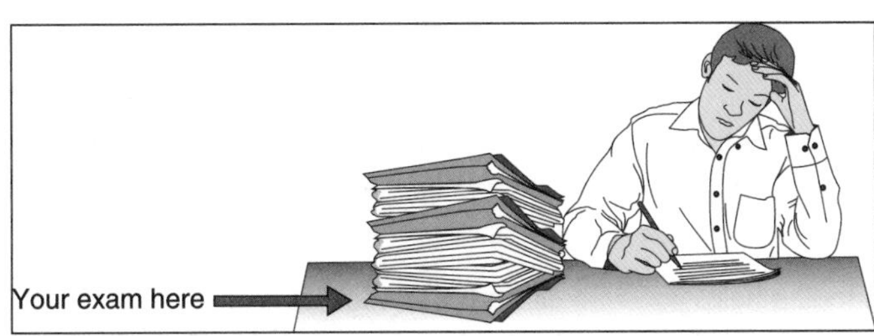

Make his or her life a bit easier and you may just get a mark here or there as a result. One really nice thing you can do on Paper 2 is write your answers vertically, leave lots of space and only answer one question per page.

1. (a) $x = 6$ 2. $2x + 5 = 10$ (b) $y = 2x + 5 = 2 \times 6 + 5 = 17$ (c) $\frac{-b}{2a} = \frac{-12}{6} = -2$ 3. (a) $p \vee q$ 4. (a) $\frac{4}{5}$ (b) If Marty has no (b) $\frac{5}{12} + \frac{8}{20}$ brains then he will $= \frac{13}{32}$ not get a girlfriend. (c) $\frac{100}{123}$ (c) If Marty will not (d) $\frac{3}{4} \times 7/9 =$ get a girlfriend then $\frac{21}{36}$ he has no brains. (d) it is a tautology	1. (a) $x = 6$ (b) $y = 2x + 5$ $= 2 \times 6 + 5$ $= 17$ (c) $\frac{-b}{2a}$ $= \frac{-12}{6}$ $= -2$ [you would then start a new page and go on]
bad paper	good paper

Hint #3: Don't use calculator notation

This is a big no-no. You're not allowed to write things like "I did 2nd CALC minimum and found (–1E20, 3E4)". Make sure you use proper maths notation whenever you write on an IB exam.

Hint #4: Get your follow-through marks

Many students "give up" on a question. Don't! If you're trying to do part (a) and just can't get it, write a wrong *but reasonable* answer and use it to do parts (b), (c), (d), etc. You'd be amazed how many marks you can get for wrong answers with the correct method!

Hint #5: Read the problem

Read the problem. Read the problem. Read the problem. Read the problem. Get it? Read the problem! I can't tell you the number of marks I take off because the student forgot to write "$x =$", forgot to round to 4dp, forgot that the answer was supposed to be in euros, etc.

Hints from Hana and Chris, the IB students

Hi! We're students who recently made it through the IB math programme. We learned over the course of our two years that there are things that really make maths exams easier. We figure it's better for you to hear it from us than spend time worrying about it when you could be studying for other things – or not! – so here are a few of the tips we learned the hard way.

Exam tips

1. Know your calculator.

   This is one of the most important things you can do when you take your exams. If you know from practice in which situations your calculator is helpful, and ways to use it efficiently (PolySmlt App with the root finder is great!), you won't be spending time figuring out your calculator during the exam – time that you can use to solve the problems instead.

2. Know how to use the help button.

   The Catalog Help application is a wonderful tool. It helps you with all kinds of things, particularly in Statistics where notation on the calculator and the arrangement of upper and lower bounds, means, and standard deviations can be tricky. Use the help button particularly in statistics problems to make sure all your information is put into the calculator correctly (and in the correct order!).

3. Do what you can in your head.

   Despite the face that the calculator is a great tool, it can hinder speed on the exam. Learn how to do the basic arithmetic in your head, so you're not doing basic subtraction like 22 – 11 on your calculator. Contrary to popular belief, the calculator isn't always faster. There's no substitute for knowing your times tables.

4. Be familiar with notation, both on the calculator and on the papers.

   There are two kinds of notation: IB notation and calculator notation. Be sure that you don't use calculator notation inappropriately – it's a silly penalty, but the IB examiners will take marks off if they see improper notation, even if your maths is correct and you've obviously done the problems right. However, you also need to understand the notation used by the calculator; sometimes you just can't put IB notation into the calculator. Be able to switch back and forth with ease and you'll avoid losing easy marks on the final paper.

5. Watch your time.

   One of the biggest problems during both studying and exam-taking is time management. When you're studying, make sure you pay as much attention to the topics you don't know as to the topics you do know. That way, when the exam comes, you'll know them both equally. Also be sure to watch your time during the exam – if you're stuck on one problem, go on to the next. It's the best way to get as many marks as you can. Remember, the problems are generally weighted the same, so the more you can do, the better.

6. Get the easy marks.

   One of the best strategies is to use the reading period effectively. Go through the test and find the problems you can easily answer. Do those first – they'll be easy marks. Then go through and get as many method marks as you can; save the problems you find difficult for the end. After all, the harder problems aren't necessarily worth more marks than the easy ones!

GOOD LUCK!

Hints from Scott, the IB teacher

Hint #1: Set up your GDC for best results!

Most students I've encountered don't realise how important it is to set up their GDC properly before the exam. If you do just a few things, it will help you a lot!!

The IB allows you to have a few helpful applications on your GDC (called Apps) – PolySmlt can be especially useful.

Also, you should make sure that your calculator is set to degrees (for trigonometry) and that the diagnostics are turned on (for statistics). Here's how you do all this:

To get PolySmlt:

Go to the APPS menu. Scroll down and see if you have PolySmlt in the list. If you have a lot of Apps, you need to delete them for the exam anyway (your IB Coordinator will not allow you to have the Apps for the exam). If you do not have PolySmlt, you need to go to www.ti.com/calc and download it (or get it from your teacher). Then install it and practise using it!

To set your calculator to degrees:

Go to the MODE menu. Move your cursor until the word 'degree' is blinking and press ENTER. The word 'degree' should now be black.

To turn on diagnostics:

Go to the catalog screen 2nd 0. Go down a bit until you see 'DiagnosticOn'.
Press ENTER to get it onto the main screen, and then press ENTER again to do it.
It should say 'Done'.

Emergency operations

To reset your list editor (lots of students 'lose' L1, for example):

Go to STAT 5 and then press ENTER.

To reset your calculator entirely (only as a last resort!):

Go to 2nd MEM 2nd + and then go to Reset 7. Go over to ALL ▶ ▶ and then press ENTER. Press 2 and the calculator should be reset back to its factory settings. *Don't forget to now go back and turn on degrees and diagnostics (see above). You will have also lost your Apps so you should do this only if all else fails!*

Finally, don't forget to put fresh batteries into your GDC before the exam and bring extra batteries with you. Your GDC takes four AAA batteries – get good ones (Duracell or Energizer). Remember Murphy's Law: "Anything that can go wrong, will go wrong at the worst possible time." That means your GDC will die right at the beginning of Paper 1.
Don't let it happen to you!

Hint #2: Don't forget the basics

IB Maths Studies is a great course for many reasons – my favourite one is that the course is *accessible*. I tell students every year that anyone can get through this course. But the problem most students have with Maths Studies has nothing to do with Maths Studies. They have problems with maths itself! If they loved mathematics and were mega-geniuses in mathematics, they would probably be taking Maths HL. (Am I right?) So this course is made for **you** – the fantastic, super-creative person who may have had trouble learning the basics when you had that weird old creepy maths teacher four years ago. Well the basics have not gone away and they are the bane of all Maths Studies students. Make sure you review things like:

- how to work with fractions
- how to work with exponents
- how to work with basic equations
- how to work with formulae
- how to graph a line
- the order of operations (PEMDAS or something like that)
- basic geometry, such as area and perimeter
- Pythagoras' Theorem
- ratios and proportions.

There's a section in your teacher's syllabus called "Presumed Knowledge". Get a copy of that and make sure you know everything on it. Anything in Presumed Knowledge is allowed to be on an IB exam!

Hint #3: Notation, notation, notation – oh how I love you!

Things like this drive a Maths Studies student crazy: "write your answer in the form $a \times 10^k$ where $k \in \mathbb{Z}$ and $1 \leq a < 10$". Why didn't they just say, "write your answer in scientific notation"? It's a long story but it's really important that you learn (and love?) maths notation, or at least how it relates to your IB Maths Studies exam. A lot of the notation is already in this book in various places and there's a great list in your syllabus document if you can get your hands on that, but here's a short list of the really important ones you should know.

$\in$	an element of
$a^{\frac{1}{n}} = \sqrt[n]{a}$	the *n*th root of a
$\{x \mid \ \}$	the set of all x such that
$a^{-n} - \dfrac{1}{a^n}$	reciprocal of a^n
$\approx$	approximately
$\sum_{i=1}^{n} u_i = u_1 + u_2 + \ldots + u_n$	sum formula*
$f(x) =$ or $f : x \mapsto$	function notation
$A\hat{B}C$	angle with B as vertex
$\hat{B}$	angle B
[AB]	line segment AB
AB	length of segment AB

* this is not examined but you'll see it in your formula booklet!

Also, note the difference between these words:

draw means make an accurate diagram or graph on graph paper, neatly and accurately with a ruler, etc. You label *everything*.

sketch means make a decent but not perfect diagram or graph, usually on regular paper. You only indicate intercepts, some sense of scale, axes, etc.

Many students make too-nice sketches when only regular sketches are required, and others make sloppy sketches when nice 'drawings' are required. Read carefully!

TOPIC 1 Sequences and series – arithmetic

What do you need to know?

An arithmetic sequence is a series of numbers that go up or down by the same amount each time. This same amount that you *add* or *subtract* is called d, the *common difference*.

How do you do it?

Step 1: Make sure that it's an arithmetic sequence

If it goes up or down by d each time, it's an arithmetic sequence: 7, 10, 13, 16,19, 22, …
If it doesn't do this, it's not an arithmetic sequence: 7, 9, 13, 20, 32, …

Step 2: Go to your formula booklet and get the formulae

2.5	The nth term of an arithmetic sequence	$u_n = u_1 + (n - 1)\,d$
	The sum of n terms of an arithmetic sequence	$S_n = \frac{n}{2}(2u_1 + (n-1)\,d) = \frac{n}{2}(u_1 + u_n)$

SCOTT SAYS:

Be careful – there are two versions of the formula for the sum! You only use the one that is easiest for what you are doing:

Use this one if you know the last term: $S_n = \frac{n}{2}(u_1 + u_n)$

Use this one if you don't know the last term: $S_n = \frac{n}{2}(2u_1 + (n-1)\,d)$

Step 3: Work out what you know and what is wanted

u_n = last term in sequence

u_1 = first term in sequence

n = number of terms in sequence

d = common difference

S_n = sum of the series

Step 4: Substitute and solve for what is wanted using algebra

Examples

- 3, 7, 11, 15, 19 ⇒ goes up by 4 each time – "common difference" is 4 ⇒ $d = 4$.
  To find the 32nd term, you would use the first formula:

  $u_{32} = 3+(32-1).4=3+124=127.$

  To find the sum of the first 32 terms of the series, you would do

  $S_{32} = \frac{32}{2}(3 + 127) = 2210.$
- 30, 22, 14, 6, – 2 ⇒ goes down by 8 each time – "common difference" is – 8 ⇒ $d = -8$.

To find the sum of the first 19 terms of the series without knowing the 19th term exactly, you would do.

$S_{19} = \frac{19}{2}(2 \cdot 30 + (19 - 1) \cdot -8) = 9.5(60 - 144) = 9.5 \cdot -84 = -798.$

- If a sequence has a 5th term = 17 and a common difference of 6, you would use the first formula to find the 1st term: $17 = u_1 + (17 - 1).6 \Rightarrow 17u_1 + 96 \Rightarrow u_1 = -79$.

How can my calculator help me do this?

There are two things you can do to make your life easier.
Let's create one of the examples above using our GDC:

1st method

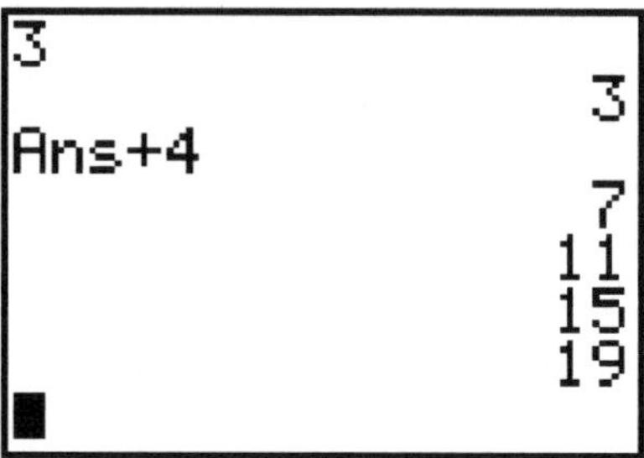

2nd method (trickier but great if you can master it!)

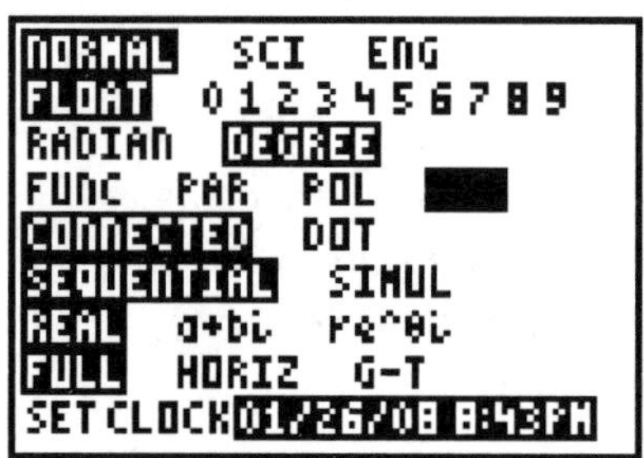

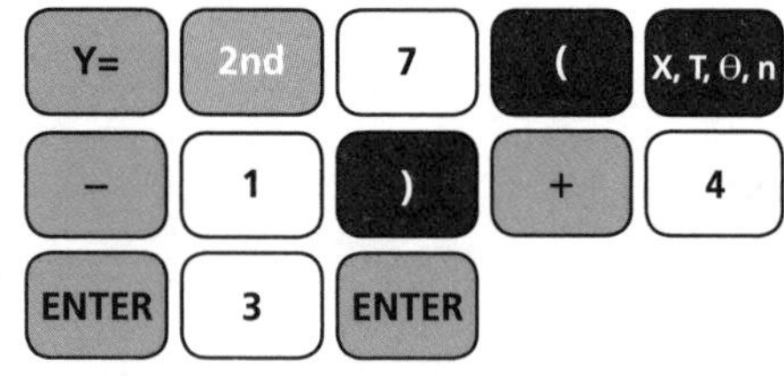

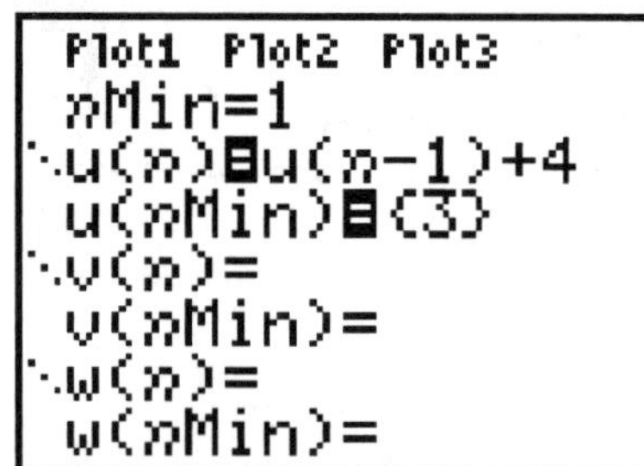

nMin means starting counting with the 1st term

u(n–1) means get the previous term

u(nMin) is the same as u_1 in our formula

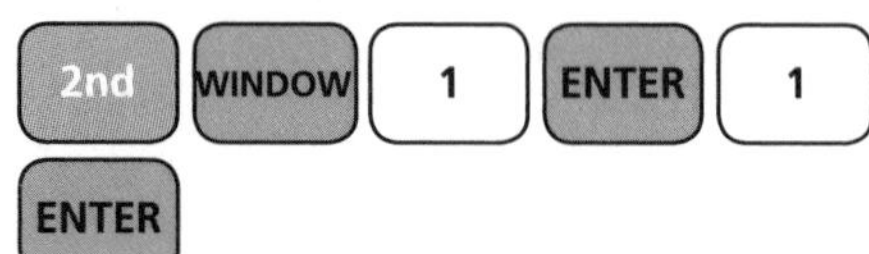

TblStart is what term you want the calculator to start showing

ΔTbl is how much to go up by (usually 1 with these problems)

2nd GRAPH

n	u(n)
1	3
2	7
3	11
4	15
5	19
6	23
7	27

n=1

Examples from past IB papers

An easy one from May 1996 Paper 1

The cost of boring a well 300 metres deep is calculated from the following information:

The cost for the first metre is \$20.00, and then the cost per metre increases by \$2.00 for every subsequent metre.

Find

(a) the cost of boring the 300th metre;

(b) the total cost of boring the well.

Working:

(a) $u_1 = 20$

$u_2 = 22$

$u_{300} = 20 + (300 - 1)(2)$

$= 618$

(b) $S_{300} = \frac{300}{2}(20 + 618)$

$= 95\ 700$

Answers:

(a) \$618

(b) \$95 700

A harder one from May 1998 Paper 1

A bank clerk is offered a yearly salary of CHF18 000 (Swiss Francs) for the first year of employment. She is also to receive CHF500 a year increase at the beginning of each subsequent year.

(a) What is her salary at the beginning of the fifth year?

After the first five years she is to receive a yearly increase of CHF750 until she reaches a maximum annual salary of CHF26 000.

(b) What is her salary at the beginning of the sixth year?

(c) At the beginning of which year of her employment does her salary reach the maximum?

Working:

(a) 18 000 + 500(4) = 20 000

(b) 20 000 + 750 = 20 750

(c) $\frac{(26\,000 - 20\,750)}{750} = 7$

6 + 7 = 13

Answers:

(a) CHF 20 000

(b) CHF 20 750

(c) 13th year

Now you practise it

An easy one from May 1999

A woman deposits $100 into her son's savings account on his first birthday. On his second birthday she deposits $125, $150 on his third birthday, and so on.

(a) How much money would she deposit into her son's account on his 17th birthday?

(b) How much in total would she have deposited after her son's 17th birthday?

Working:

Answers:

(a) ....

(b) ....

A harder one from November 2003

The fourth term of an arithmetic sequence is 12 and the tenth term is 42.

(a) Given that the first term is u_1 and the common difference is d, write down two equations in u_1 and d that satisfy this information.

(b) Solve the equations to find the values of u_1 and d.

Working:

Answers:

(a) ....

(b) ....

IRENE SAYS:

TOPIC 2

Sequences and series – geometric

What do you need to know?

A geometric sequence is a series of numbers that go up or down by the same rate each time. This same amount that you *multiply* or *divide* by is called r, the *common ratio*.

How do you do it?

Step 1: Make sure that it's a geometric sequence

If it goes up or down by multiplying by r each time, it's a geometric sequence: 3, 12, 48, 192, 768, ... (multiply by 4)
If it doesn't do this, it's not a geometric sequence: 3, 12, 24, 96, 192, ...

Step 2: Go to your formula booklet and get the formulae

2.6	The nth term of a geometric sequence	$u_n = u_1 r^{n-1}$
	The sum of n terms of a geometric sequence	$S_n = \dfrac{u_1(r^n - 1)}{r - 1} = \dfrac{u_1(1 - r^n)}{1 - r}, r \neq 1$

SCOTT SAYS:

Be careful – again there are two versions of the formula for the sum! But this time they are *exactly the same.* Why are there two? I have no idea. Just use the first one and don't use both!!

Always use this one: $S_n = \dfrac{u_1(r^n - 1)}{r - 1}$

Step 3: Work out what you know and what is wanted

u_n = last term in sequence

u_1 = first term in sequence

n = number of terms in sequence

r = common ratio

S_n = sum of the series

Step 4: Substitute and solve for what is wanted using algebra.

Examples

- 3, 12, 48, 192, 768 $\Rightarrow$ multiply by 4 each time – "common ratio" is 4 $\Rightarrow r = 4$.
  To find the 12th term of the sequence, you would use the first formula:
  $u_{32} = 3 \cdot 4^{12-1} = 3 \cdot 4194304 = 12582912.$
  To find the sum of the first eight terms, you would use the second formula:
  $S_8 = \dfrac{3 \cdot (4^{8-1})}{4 - 1} = \dfrac{3 \cdot 65535}{3} = 65535.$
- 40, 8, 1.6, 0.32, 0.064 $\Rightarrow$ multiply by 0.2 each time – "common ratio" is 0.2 $\Rightarrow r = 0.2$.
  To find the sum of the first eleven terms of the series, you would use the second formula again:
  $S_{11} = \dfrac{40 \cdot (0.2^{11-1})}{0.2 - 1} = \dfrac{40 \cdot -0.99999997952}{-0.8} = 49.999998976 \approx 50.0 \text{ (3sf)}$

How can my calculator help me do this?

There are two things you can do to make your life easier.

Let's create one of the examples above using our GDC:

SCOTT SAYS:

Be careful – in that last example, you might want to divide by 5 each time instead of multiplying by 0.2. It's the same thing but with geometric sequences we like to multiply in order to find the "common ratio".

1st method

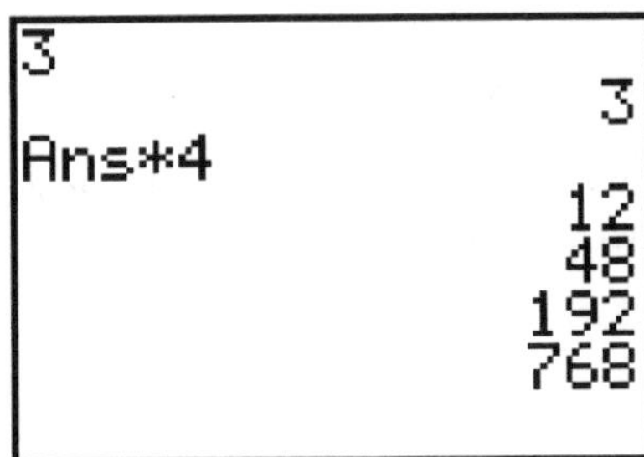

2nd method (trickier but great if you can master it!)

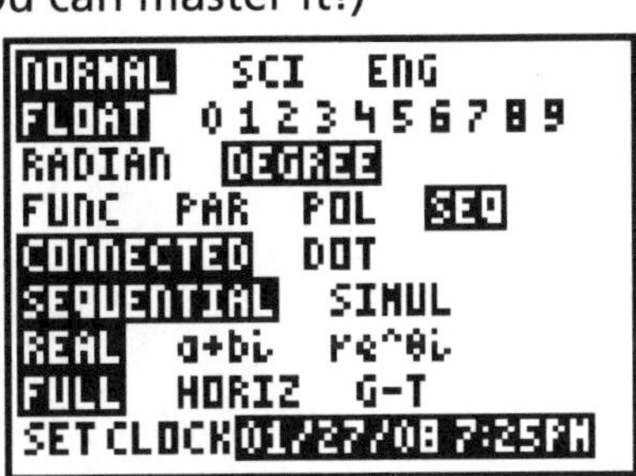

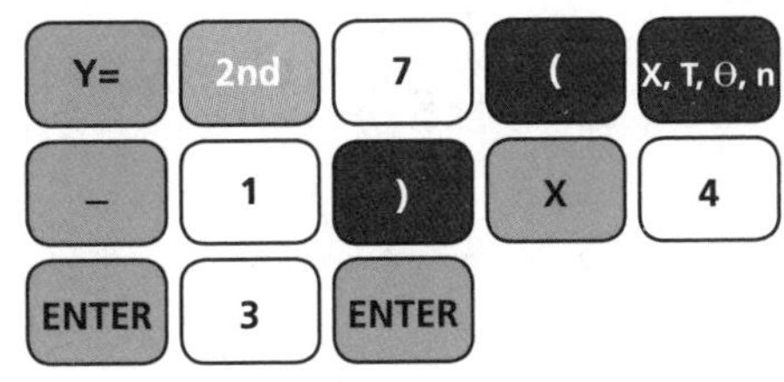

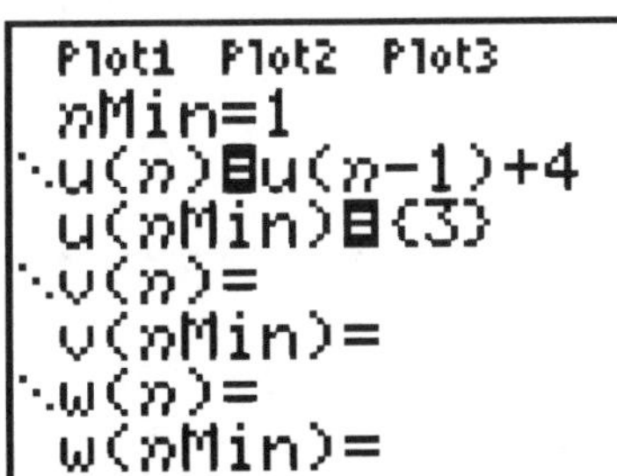

nMin means starting counting with the 1st term

u(n−1) means get the previous term

u(nMin) is the same as u_1 in our formula

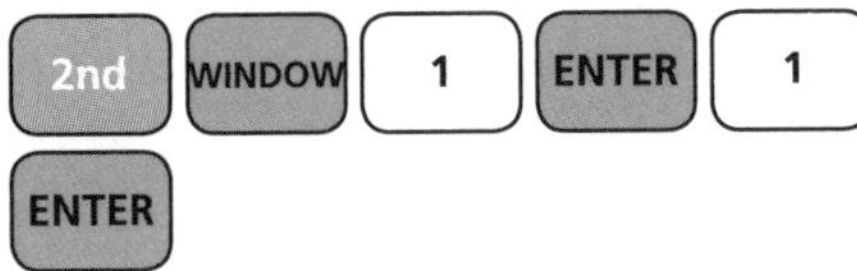

TblStart is what term you want the calculator to start showing

ΔTbl is how much to go up by (usually 1 with these problems)

2nd GRAPH

n	u(n)
1	3
2	12
3	48
4	192
5	768
6	3072
7	12288

n=1

Examples from past IB papers

An easy one from Specimen 2000 Paper 1

The tuition fees for the first three years of high school are given in the table below.

Year	Tuition fees (in dollars)
1	2 000
2	2 500
3	3 125

These tuition fees form a geometric sequence.

(a) Find the common ratio, r, for this sequence.

(b) If fees continue to rise at the same rate, calculate (to the nearest dollar) the total cost of tuition fees for the first six years of high school.

Working:

(a) $r = \frac{2500}{2000} = 1.25$

(b) $S_n = \frac{u_1(r^n - 1)}{r - 1}$

$S_6 = \frac{2000(1.25^6 - 1)}{1.25 - 1}$

$= 22\ 517.57813$

Answers:

(a) 1.25

(b) $22 518

A harder one from May 2004 Paper 2

A basketball is dropped vertically. It reaches a height of 2 m on the first bounce. The height of each subsequent bounce is 90% of the previous bounce.

(a) What height does it reach on the 8th bounce?

(b) What is the total vertical distance travelled by the ball between the first and the sixth time the ball hits the ground?

Working:

(a) $u_n = 2(0.9)^7 = 0.9565938$

(b) $S_n = \frac{2((0.9)^5 - 1)}{(0.9 - 1)} = 8.1902$

Total distance travelled = 2×8.1902

$= 16.3804$ m

Answers:

(a) 0.957 m

(b) 16.4 m

Now you practise it

An easy one from Specimen 2005 Paper 1

A geometric sequence has all its terms positive. The first term is 7 and the third term is 28.

(a) Find the common ratio.

(b) Find the sum of the first 14 terms.

Working:

Answers:

(a) ....

(b) ....

A harder one from May 2000

The population of Bangor is growing each year. At the end of 1996, the population was 40 000. At the end of 1998, the population was 44 100. Assuming that these annual figures follow a geometric progression, calculate

(a) the population of Bangor at the end of 1997;

(b) the population of Bangor at the end of 1992.

Working:

Answers:

(a) ....

(b) ....

HANA AND CHRIS SAY:

A good trick here is to make u_1 represent the year 1992 and go up from there!

TOPIC 3

Number sets

What do you need to know?

- You are required to memorise names and definitions of the various number sets:

  $\mathbb{N}$ = natural numbers = 0, 1, 2, 3, 4, 5, …

  $\mathbb{Z}$ = integers = …, −4, −3, −2, −1, 0, 1, 2, 3, 4, 5, … (positive whole numbers)

  $\mathbb{Q}$ = rational numbers = $\frac{-1}{2}, \frac{3}{4}, \frac{-32}{9}, \frac{7}{1}$, … (numbers that can be written as a fraction)

  $\overline{\mathbb{Q}}$ = irrational numbers = $\sqrt{2}$, π, $\sqrt{31}$, …

  $\mathbb{R}$ = real numbers = everything!

  (There are actually numbers that are not "real" – they're called "imaginary" – but don't worry about them. They're not in the exam!)

- and you should also know:

  prime number = integer greater than 1 that is only divisible by 1 and itself
  = 2, 3, 5, 7, 11, 13, 17, …

SCOTT SAYS:

Students often mix up $\mathbb{Z}$ and $\mathbb{Q}$. Here's a good way to remember the difference: Q stands for *quotient* meaning division, so that's the one for fractions. Don't forget that an integer and a natural number can both be written as a fraction!

How do you do it?

Usually the exam just asks you to place numbers in the various sets.

Examples

- 6 is a natural number, an integer, a rational number and a real number
- $\frac{3}{4}$ is a rational number and a real number
- π is an irrational number and a real number
- $\sqrt{4}$ is a natural number, an integer, a rational number and a real number (because it is equal to 2)
- −192 is an integer, a rational number and a real number

How can my calculator help me do this?

There really isn't much your calculator can do here except check square roots to see if they are natural numbers.

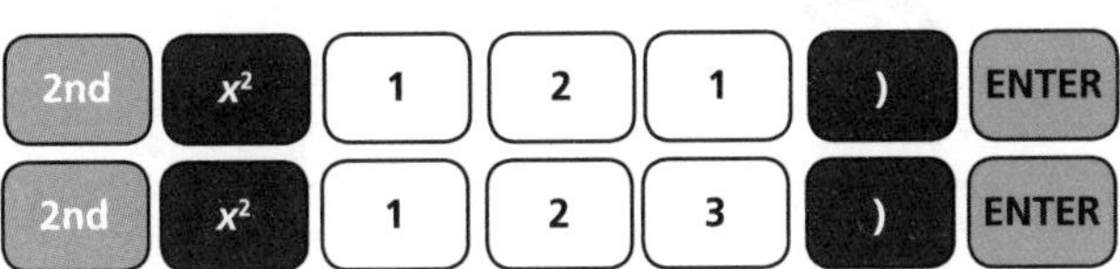


```
√(121)
                11
√(123)
       11.09053651
```

Examples from past IB papers

An easy one from May 2004 Paper 1

Let $U = \left\{-4, -\frac{2}{3}, 1, \pi, 13, 26.7, 69, 10^{33}\right\}$.

A is the set of all the integers in U.

B is the set of all the rational numbers in U.

(a) List all the prime numbers contained in U.

(b) List all the members of A.

(c) List all the members of B.

(d) List all the members of the set $A \cap B$.

Working:

Answers:

(a) $\{13\}$

(b) $A = \{-4, 1, 13, 69, 10^{33}\}$

(c) $B = \{-4, -\frac{2}{3}, 1, 13, 26.7,$ $69, 10^{33}\}$

(d) $A \cap B = \{-4, 1, 13,$ $69, 10^{33}\}$

SCOTT SAYS:

The last part of this question also uses the idea of intersection of sets which is coming up in the next chapter. Sorry! Did you remember how to do it?

A harder one from May 2005 Paper 1

(a) Given $x = 2.6 \times 10^4$ and $y = 5.0 \times 10^{-8}$, calculate the value of $w = x \times y$. Give your answer in the form $a \times 10^k$ where $1 \leq a < 10$ and $k \in \mathbb{Z}$.

(b) Which two of the following statements about the nature of x, y and w are **incorrect**?

(i) $x \in \mathbb{N}$

(ii) $y \in \mathbb{Z}$

(iii) $y \in \mathbb{Q}$

(iv) $w < y$

(v) $x + y \in \mathbb{R}$

(vi) $\frac{1}{w} < x$

Working:

(a) $w = (2.6 \times 10^4) \times (5.0 \times 10^{-8})$

$= 13 \times 10^{-4}$ (or 0.0013)

$= 1.3 \times 10^{-3}$

Answers:

(a) 1.3×10^{-3}

(b) (ii) and (iv) are incorrect

SCOTT SAYS:

The first part of this question also uses the idea of scientific notation which is coming up in a later session. Sorry! Did you remember how to do it?

Now you practise it

An easy one from November 2002 Paper 1

Consider the numbers 5, 0.5, $\sqrt{5}$ and -5. Complete the table on the next page, showing which of the number sets $\mathbb{N}$, $\mathbb{R}$ and $\mathbb{Q}$ these numbers belong to.

Working:

Answers:

	$\mathbb{N}$	$\mathbb{R}$	$\mathbb{Q}$
5			✓
0.5	×		
$\sqrt{5}$	×		
−5		✓	

IRENE SAYS:

Don't worry about that working box! I have to give you a box to work in, even if you don't use it. Just leave it blank for problems like this.

A harder one from Specimen 2005 Paper 1

Given $\mathbb{Z}$ the set of integers, $\mathbb{Q}$ the set of rational numbers, $\mathbb{R}$ the set of real numbers.

(a) Write down an element that belongs to $\mathbb{R} \cap \mathbb{Z}$.

(b) Write down an element that belongs to $\mathbb{Q} \cap \mathbb{Z}'$.

(c) Write down an element that belongs to $\mathbb{Q}'$.

(d) Use a Venn diagram to represent the sets $\mathbb{Z}$, $\mathbb{Q}$ and $\mathbb{R}$.

Working:

Answers:

(a) ....

(b) ....

(c) ....

(d) ....

HANA AND CHRIS SAY:

If you forgot all those set symbols, go to the sets section and come back to this one.

TOPIC 4

Approximation, significant figures, percentage errors, estimation

What do you need to know?

You need to be able to round numbers to a particular stated precision (decimal places or dp), round numbers to a particular number of significant figures or sf (usually three), calculate a percent error of a measurement or calculation and estimate a measurement or calculation.

IRENE SAYS:

Don't forget that all answers on the whole exam must be rounded to 3 significant figures unless otherwise stated. If you forget, there's a one-mark "accuracy penalty" per paper.

How do you do it?

For approximation and estimation, you just use your common sense. For example, if a measurement is accurate to the nearest cm, then all answers are ± 0.5 cm.

For significant figures, each non-zero digit is "significant" and zeros are significant only at the end of an integer or in between non-zero digits. We usually abbreviate significant figures to "sf".

For percentage error, you go to your formula booklet and get the formula.

2.2	Percentage error	$v = \left\|\frac{v_E - v_A}{v_E}\right\|$, where v_E is the exact value and v_A is the approximate value of v

Examples

- If you have a rectangle with sides 5 cm and 3 cm in length, rounded to the nearest cm, then the smallest area it can have is 4.5 cm × 2.5 cm = 11.25 cm^2
- If a calculation answer is 9034.204 821, then the answer is 9000 to 1sf, 9000 to 2sf, 9030 to 3sf, 9034 to 4sf, 9034.2 to 5 sf, 9034.20 to 6 sf, 9034.205 to 7sf, etc.
- If a calculation answer is 12.7251934, then the answer is 12.7 to 1dp, 12.73 to 2dp, 12.725 to 3dp, 12.7252 to 4dp, etc.
- If you measure a circle and get an estimate of π that is 3.2, the percentage error would be $\left|\frac{\pi - 3.2}{\pi}\right| = |-0.01859| = 0.01859 = 1.86\%$ (3sf)

How can my calculator help me do this?

Your calculator can round but it cannot do significant figures. It can also help you to do percentage errors.

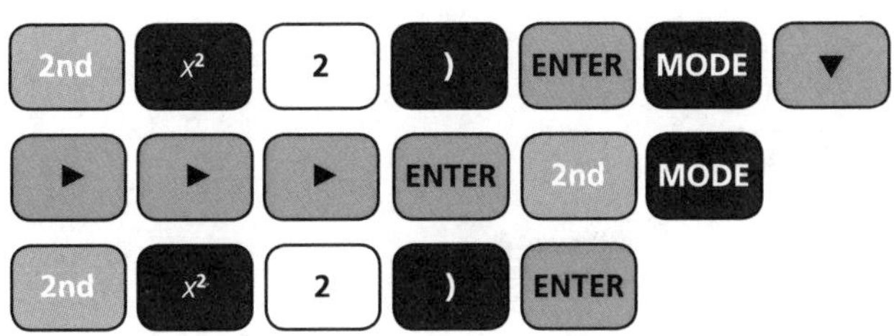

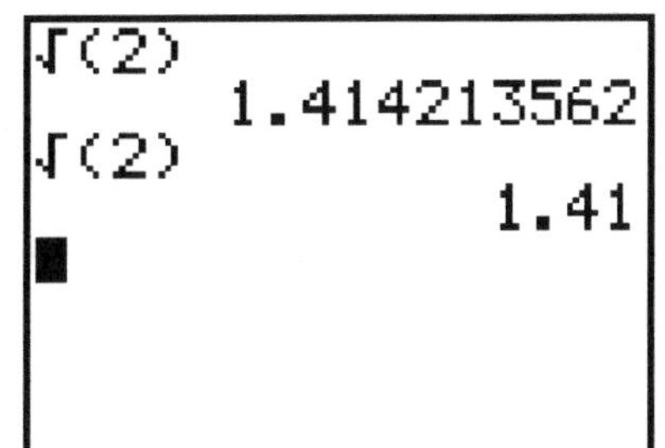

**Warning – don't forget to set your calculator back to "float" when you're done with this rounding feature!! **

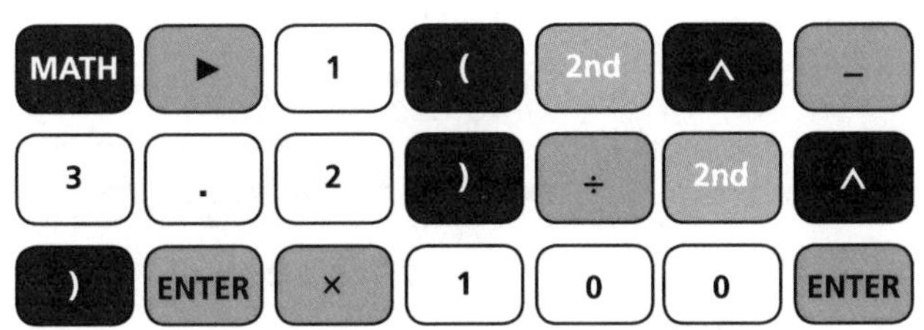

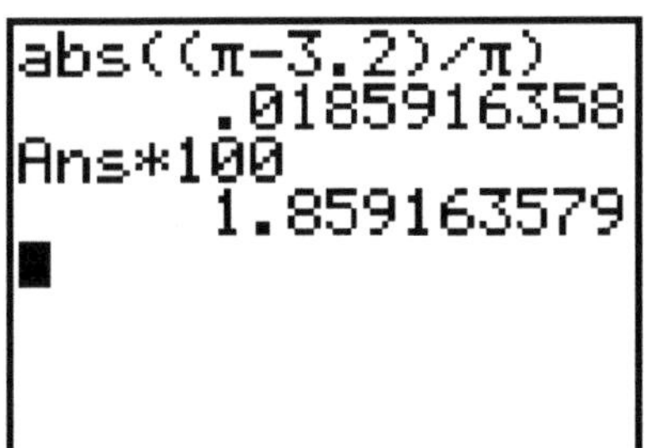

abs means absolute value – those bars around the percentage error formula.

Examples from past IB papers

An easy one from May 1999 Paper 1

Calculate $3.7 \times 16.2^2 - 500$, writing your answer

(a) correct to two decimal places;

(b) correct to three significant figures.

Working:

(a) $3.7 \times 16.2^2 - 500 = 471.028 \rightarrow 471.03$ (2 decimal places)

(b) $471.028 \rightarrow 471$ (3 significant figures)

Answers:

(a) 471.03

(b) 471

A harder one from November 2000 Paper 1

Anthony uses the formula

$$p = \frac{27q}{r+s}$$

to calculate the value of p when, correct to two decimal places, $q = 0.89$, $r = 1.87$ and $s = 7.22$.

(a) He estimates the value **without using a calculator.**

(i) Write down the numbers Anthony could use in the formula to estimate the value of p.
(ii) Work out the estimate for the value of p that your numbers would give.

(b) A calculator is to be used to work out the actual value of p.
To what degree of accuracy would you give your calculator answer? Give a reason for your answer.

Working:

(a)(ii) $p = \frac{27(1)}{(2 + 7)} = 3$

Answers:

(a) (i) $q = 1, r = 2, s = 7$

(ii) $p = 3$

(b) Two decimal places, because the variables are given to two decimal places.

Now you practise it

An easy one from November 2003 Paper 1

Using the formula $V = \pi r^2 (H - h)$, and your calculator value for π, calculate the value of V when $r = 4.26$, $H = 21.58$ and $h = 14.35$.

(a) Give the full calculator display.

(b) Give your answer to two decimal places.

(c) Give your answer to two significant figures.

Working:

Answers:

(a) ....

(b) ....

(c) ....

A harder one from May 2004 Paper 1

Arthur needs to calculate a value from a trigonometric formula. He uses his calculator to find the value of r given by $r = \dfrac{1}{\sin(86°) - \sin(85°)}$.

(a) Calculate the value of r, correct to three significant figures.

(b) Arthur makes the mistake of rounding both of the sines to three significant figures **before** taking their difference. Calculate the value of r found by Arthur. Call this value r_A.

(c) Calculate the percentage error E in Arthur's calculation.

Working:

Answers:

(a) ....

(b) ....

(c) ....

TOPIC 5

Scientific notation and the SI (metric) system

What do you need to know?

In scientific notation, you put the number in the form $a \times 10^k$, where $1 \leq a < 10$, $k \in Z$.

In the SI system, you need to know the basic values first: length in metres (m), volume in litres (L), mass in grams (g), speed in metres per second ($m\ s^{-1}$), acceleration in metres per second squared ($m\ s^{-2}$), time in seconds (s) and so on. The basic prefixes are: milli ($\frac{1}{1000}$), centi ($\frac{1}{100}$), deci ($\frac{1}{10}$), deka (10) and kilo (1000). There are others but these are the usual ones.

How do you do it?

To put a number into scientific notation, you move the decimal point to the left or right until there is only one non-zero digit to the left of the point. You count the number of places you moved and give that value to k. If you move to the left, k is positive. If you move to the right, k is negative.

To convert SI units, you either multiply or divide by the relative value of the conversion, for example, if you want to convert $m\ s^{-2}$ to $km\ s^{-2}$, you know that there are 1000 metres in a kilometre and $km\ s^{-2}$ is going to be smaller than $m\ s^{-2}$, so you would divide by 1000.

Examples

- To convert the number 9034.204 821 to scientific notation, you move the decimal point three places to the left. After rounding to 3sf, the answer becomes 9.03×10^3.
- To convert the number 0.000 023 182 2 to scientific notation, you move the decimal point five places to the right. After rounding to 3sf, the answer is 2.32×10^{-5}.
- To convert 0.001 297 1 kilograms (kg) to grams (g), you multiply by 1000 and get 1.30 g (3sf).
- To convert 120 000 $cm\ s^{-1}$ to $m\ s^{-1}$, you divide by 100 and get 1200 $m\ s^{-1}$.

How can my calculator help me do this?

Your calculator can convert numbers to scientific notation, but like the rounding feature showed earlier, you must make sure you put the calculator back to "float" after you are done.

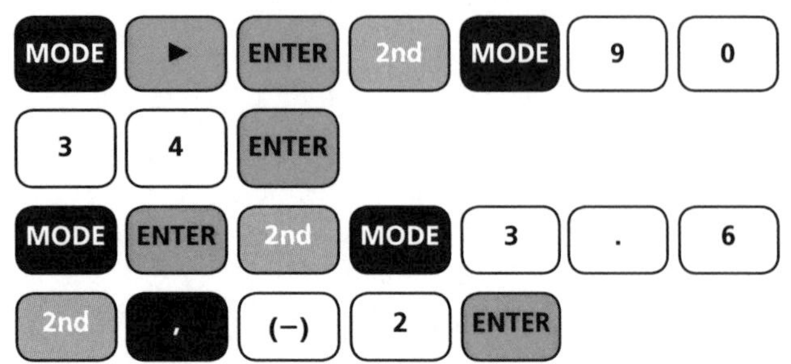

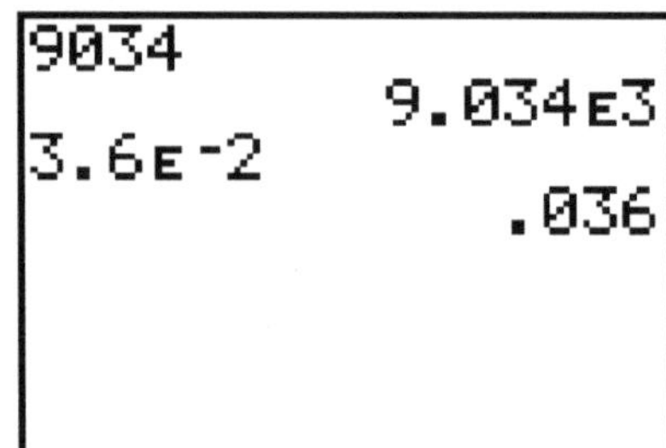

There is no feature on the GDC to convert SI units. Sorry!

IRENE SAYS:

Be careful! If you write in "calculator notation" in the exam, you will not get full marks. If you see 9.036E3, you must write 9.036×10^3.

Examples from past IB papers

An easy one from May 2000 Paper 1

Let $m = 6.0 \times 10^3$ and $n = 2.4 \times 10^{-5}$.

Express each of the following in the form $a \times 10^k$, where $1 \leq a < 10$ and $k \in \mathbb{Z}$.

(a) mn;

(b) $\frac{m}{n}$.

Working:

(a) $mn = 6.0 \times 10^3 \times 2.4 \times 10^{-5}$

$= 6.0 \times 2.4 \times 10^{-2}$

$= 14.4 \times 10^{-2}$

$= (1.44 \times 10^1) \times 10^{-2}$

$= 1.44 \times 10^{-1}$

(b) $\frac{m}{n} = \frac{(6.0 \times 10^3)}{(2.4 \times 10^{-5})}$

$= \frac{6.0}{2.4} \times 10^8$

$= 2.5 \times 10^8$

Answers:

(a) 1.44×10^{-1}

(b) 2.5×10^8

A harder one from Specimen 2005 Paper 1

A field is 91.4 m long and 68.5 m wide.

(a) Calculate the area of the field in m^2.

(b) Calculate the area of the field in cm^2.

(c) Express your answer to (b) in the form $a \times 10^k$, where $1 \leq a < 10$ and $k \in \mathbb{Z}$.

Working:

(a) $91.4 \times 68.5 = 6260.9\ m^2$

(b) $9140\ cm \times 6850\ cm = 62\ 609\ 000\ cm^2$

(c) 6.2609×10^7

Answers:

(a) $6260.9\ m^2$

(b) $62\ 609\ 000\ cm^2$

(c) $6.26 \times 10^7\ cm^2$

Now you practise it

An easy one from November 2000 Paper 1

Let $A = 4.5 \times 10^{-3}$ and $B = 6.2 \times 10^{-4}$. Find

(a) AB;

(b) $2(A + B)$.

Working:

Answers:

(a) ....

(b) ....

A harder one from May 2005 Paper 1

(a) Convert 0.001 673 litres to millilitres (ml). Give your answer to the nearest ml.

The SI unit for energy is joules. An object with mass *m* travelling at speed *v* has energy given by $\frac{1}{2}mv^2$ (joules).

(b) Calculate the energy of a comet of mass 351 223 kg travelling at a speed 176.334 m s^{-1}. Give your answer correct to six significant figures.

In the SI system of units, distance is measured in metres (m), mass in kilograms (kg) and time in seconds (s). The momentum of an object is given by the mass of the object multiplied by its speed.

(c) Write down the correct combination of SI units (m, kg, s) for momentum.

Working:

Answers:

(a) ....

(b) ....

(c) ....

SCOTT SAYS:

This question was a bit controversial among teachers after it came out in 2005. It's a tricky one and borderline whether or not it was on the syllabus. But you never know ... Irene the IB Examiner is known for putting questions like this on the exam! Stay alert!

TOPIC 6

Word problems and systems of linear equations

What do you need to know?

You need to know how to read, translate into algebra and solve a word problem. There are no required techniques to memorise; the IB want to know if you can look at a problem and solve it mathematically.

You also need to know how to solve a system of linear equations both using a graph and algebraically.

How do you do it?

For word problems, unfortunately there is no one method to convert a word problem into an equation. However here is a step-by-step approach that is usually helpful:

Step 1: Find out what is wanted and assign a variable to it (usually students like x but it can be anything)

Step 2: Translate the words and/or pictures into one or more equations

Step 3: Solve the equation

Step 4: Answer the question with the appropriate units if needed

To solve a system of linear equations (usually two equations with two variables like x and y), you can either draw a graph or use algebra.

If you draw a graph, you will get two lines and the point of intersection is where the solution lies.

If you want to do it algebraically, you either solve by substitution (solve one equation for a variable and put that expression into the other equation) or by elimination (add or subtract the equations so that one of the variables "is eliminated").

Example

Take a rectangle where one side is 3 cm longer than the other. If the perimeter is 30 cm, you can find the length of each side.

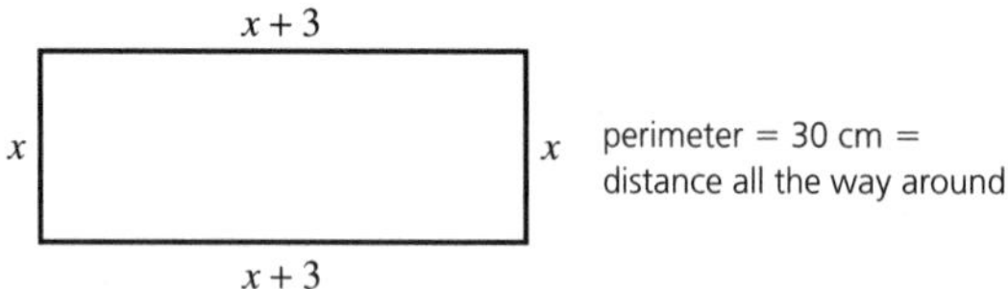

Let x = the length of the shorter side, so $x + 3$ = length of the longer side

Hence the equation is $x + (x + 3) + x + (x + 3) = 30$

Simplifying you get $4x + 6 = 30$, and if you solve you then get $x = 6$

Therefore the length of the shorter side is 6 cm and the longer side is 9 cm.

To solve the system of linear equations $\begin{cases} 2a - 3b = -14 \\ -5a + b = 9 \end{cases}$:

Graphically	By substitution	By elimination
Turn the equations into: $\begin{cases} 2x - 3y = -14 \\ -5x + y = 9 \end{cases}$ Graph the two lines and find the intersection: $(-1, 4)$ Solution is $x = -1, y = 4$ Answer is $a = -1, b = 4$	Add $5a$ to both sides of the 2nd equation to get: $b = 9 + 5a$ Substitute that into the 1st equation: $2a - 3(9 + 5a) = -14$ Solve: $2a - 27 - 15a = -14$ $-13a - 27 = -14$ $-13a = 13$ $a = -1$ Substitute to get b: $b = 9 + 5(-1) = 9 + -5 = 4$ Answer is $a = -1, b = 4$	Multiply the 2nd equation by 3: $-15a + 3b = 27$ Line up the two equations: $2a - 3b = -14$ $-15a + 3b = 27$ Add to eliminate the b's: $-13a = 13$ and solve: $a = -1$ Substitute to get b: $2(-1) - 3b = -14$ $-2 - 3b = -14$ $-3b = -12$ $b = 4$ Answer is $a = -1, b = 4$

How can my calculator help me do this?

Your calculator can solve both equations and systems of linear equations very well! This is one of the best times to use your GDC…

Method 1 to solve the linear equation $4x + 6 = 30$ – using a graph:

(Notice how you have to change the window in order to see the point of intersection)

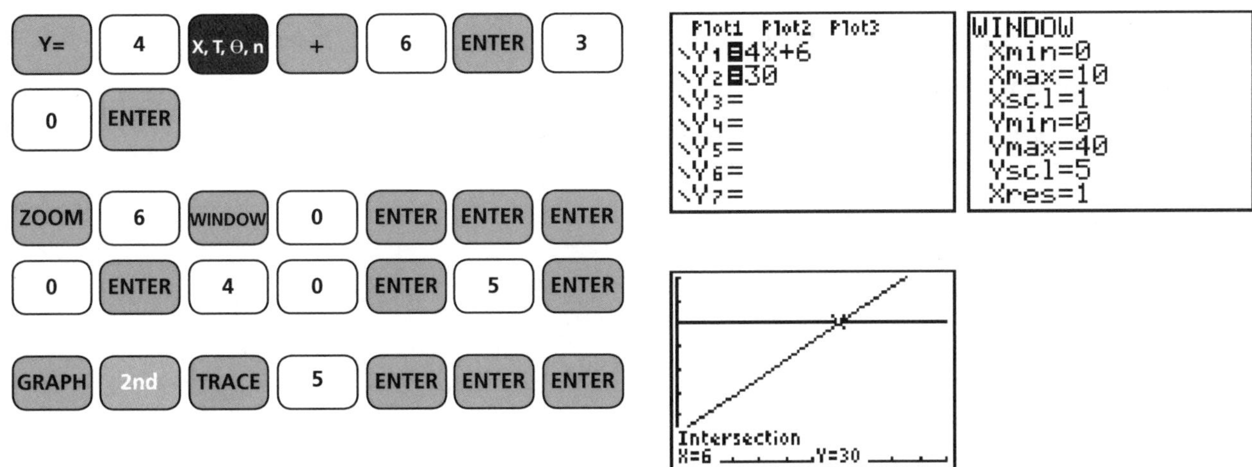

Method 2 to solve the linear equation $4x + 6 = 30$ – using the Solver:

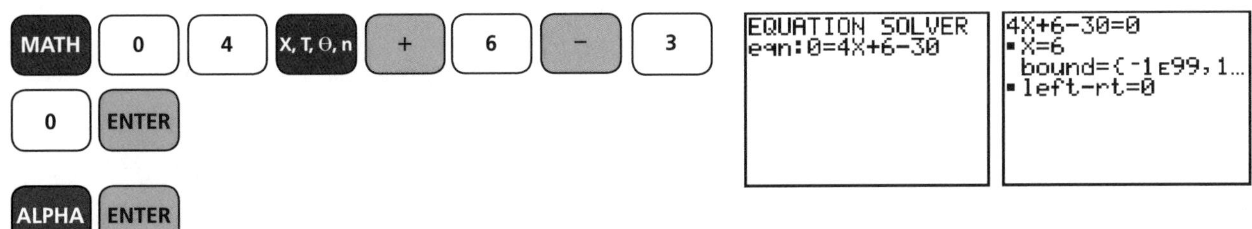

Method 1 for solving the system of linear equations $\begin{cases} 2a - 3b = -14 \\ -5a + b = 9 \end{cases}$ – using a graph:

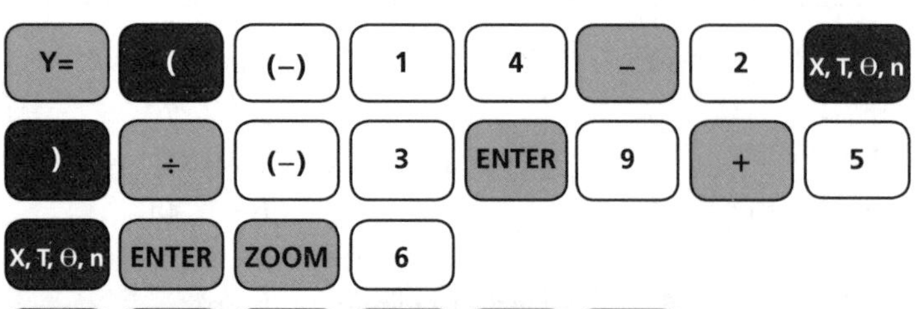

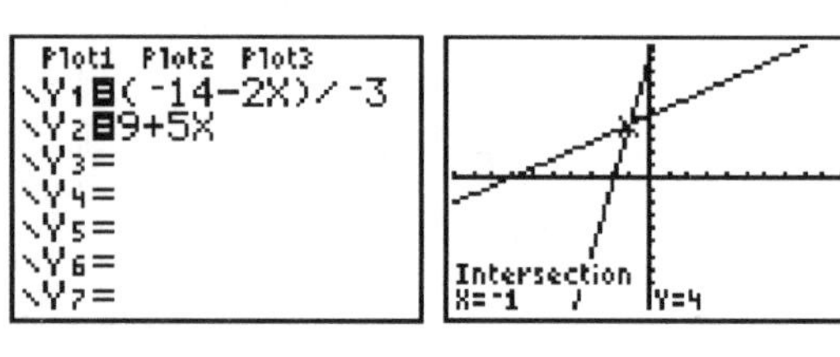

Method 2 for solving the system of linear equations $\begin{cases} 2a - 3b = -14 \\ -5a + b = 9 \end{cases}$ – using PolySmlt

(an App on your GDC that you're allowed to use in the exam!):

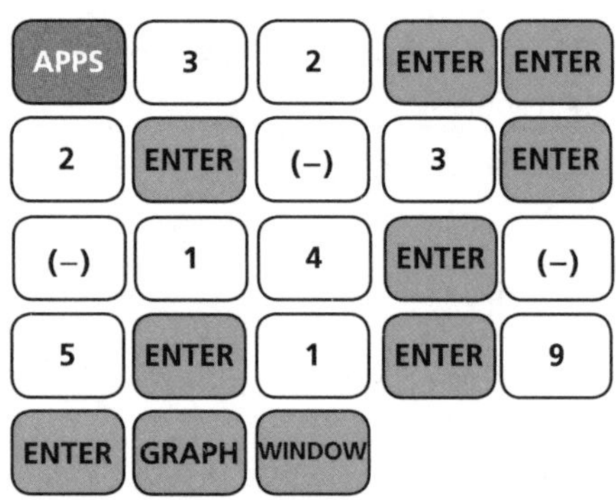

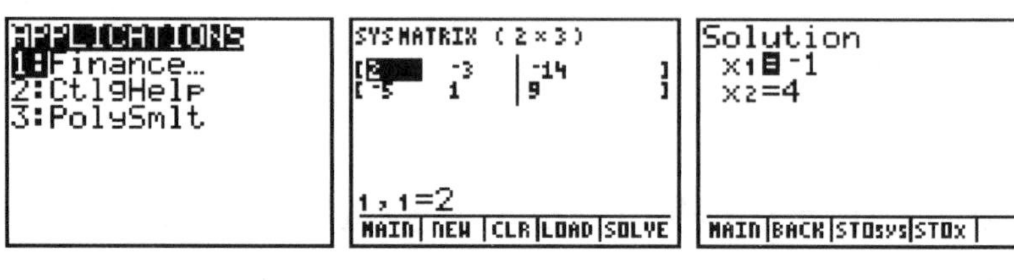

Examples from past IB papers

An easy one from May 1999 Paper 1

At Jumbo's Burger Bar, Jumbo burgers cost £J each and regular cokes cost £C each. Two Jumbo burgers and three regular cokes cost £5.95.

(a) Write an equation to show this.

(b) If one Jumbo burger costs £2.15, what is the cost, in pence, of one regular coke?

Working:

(a) $2J + 3C = 5.95$

(b) $2(2.15) + 3C = 5.95$

$4.30 + 3C = 5.95$

$3C = 1.65$

$C = 0.55$

Answers:

(a) $2J + 3C = 5.95$

(b) 55 pence

A harder one from November 2002 Paper 1

Keisha had 10 000 USD to invest. She invested m USD at the *Midland Bank,* which gave her 8% annual interest. She invested f USD at the *First National Bank,* which gave 6% annual interest. She received a total of 640 USD in interest at the end of the year.

(a) Write two equations that represent this information.

(b) Find the amount of money Keisha invested at each bank.

Working:

(a) $0.08m + 0.06f = 640$

$m + f = 10\,000$

(b) $8m + 6f = 64\,000$

$8m + 8f = 80\,000$

$2f = 16\,000$

$f = 8000$

$m = 2000$

Answers:

(a) $0.08m + 0.06f = 640$;

$m + f = 10\,000$

(b) \$8000 at First National,

\$2000 at Midland

Now you practise it

An easy one from May 2002 Paper 1

The cost c, in Australian dollars (AUD), of renting a bungalow for n weeks is given by the linear relationship $c = nr + s$, where s is the security deposit and r is the amount of rent per week.

Ana rented the bungalow for 12 weeks and paid a total of 2925 AUD.

Raquel rented the same bungalow for 20 weeks and paid a total of 4525 AUD.

Find the value of

(a) r, the rent per week;

(b) s, the security deposit.

Working:

Answers:

(a) ....

(b) ....

A harder one from May 1998 Paper 1

John pays one-fifth of his weekly earnings to his parents for housekeeping. He earns x pounds (£) per week.

(a) Write an expression, in x, which represents how much John pays his parents each week.

John receives a 7% pay rise.

(b) Write a second expression, in x, which represents how much **extra** John earns each week.

John earns £214 per week after his pay rise.

(c) How much did he earn per week before the pay rise?

Working:

Answers:

(a) ....

(b) ....

(c) ....

TOPIC 7

Quadratic equations

What do you need to know?

You need to know how to factorise a quadratic equation, how to solve a quadratic equation graphically or by factorising, and how to solve a word problem using quadratic equations.

How do you do it?

To factorise a quadratic equation, you usually have to put it into two sets of brackets so that when you multiply the brackets, you get back what you started with! The hard part is what to put where. One way is to use a box*:

To factorise $x^2 + 8x + 7$:

	x	7
x	x^2	$7x$
1	$1x$	7

so the factorisation is $(x + 1)(x + 7)$

To factorise $3x^2 + 7x - 6$:

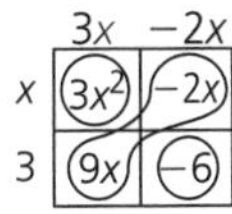

so the factorisation is $(3x - 2)(x + 3)$

To solve a quadratic equation graphically, look at the x-intercepts. These are called "roots" or "zeros". The two (or sometimes only one) points where the parabola crosses/touches the x-axis are the solutions of the equation.

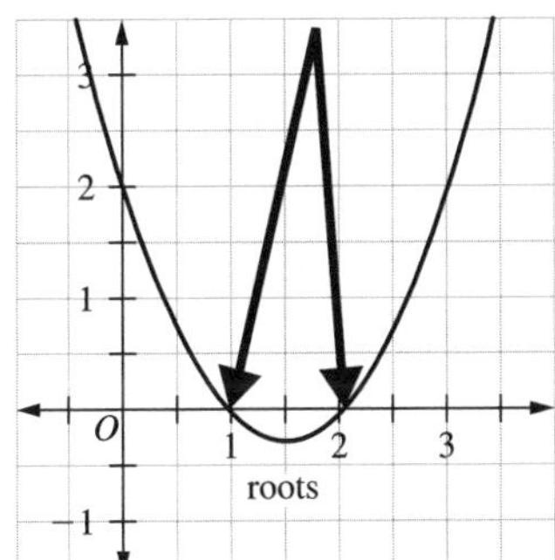

To solve a quadratic equation by factorising, you set the equation equal to zero, and factorise the quadratic expression. Then you set each factor equal to zero and solve.

Example

To solve the quadratic equation $x^2 + 3 = 4x$:

Graphically:

$y = x^2 - 4x + 3$

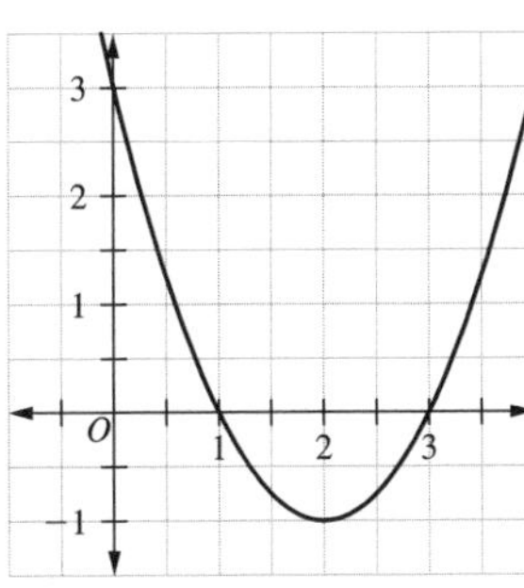

So $x = 1$ or $x = 3$

By factorising:

$x^2 - 4x + 3 = 0$

$(x - 3)(x - 1) = 0$

$x - 3 = 0$ or $x - 1 = 0$

$x = 3$ or $x = 1$

How can my calculator help me do this?

In lots of ways!! You should **definitely** use your GDC for quadratic equation problems.

Method 1 to solve the quadratic equation $x^2 + 3 = 4x$ – using a graph:

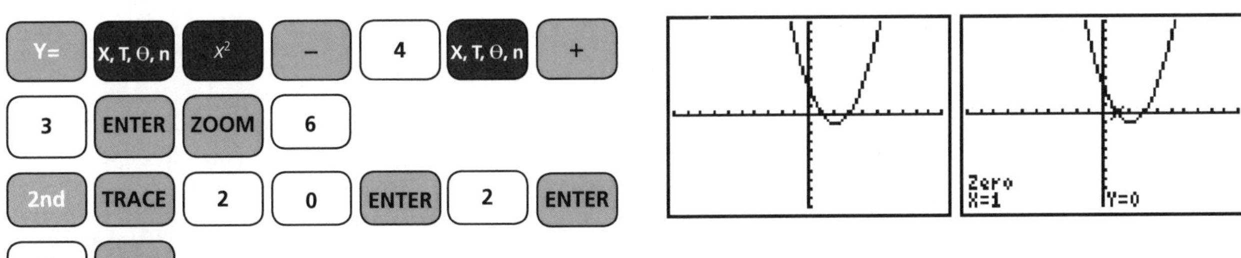

* Method adopted from Harold R. Jacobs' outstanding algebra book *Elementary Algebra*, published in 1979 by W.H. Freeman, New York, ISBN 0-7167-1047-1.

Method 2 to solve the linear equation $x^2 + 3 = 4x$ – using PolySmlt:

APPS (find PolySmlt)

ENTER 1 2 ENTER 1 ENTER (−) 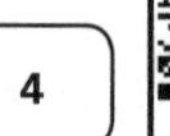 4

ENTER 3 ENTER GRAPH

APPLICATIONS
1:Finance...
2:CBL/CBR
3:CtlgHelp
4:PolySmlt

MAIN MENU
1:Poly Root Finder
2:Simult Eqn Solver
3:About
4:Poly Help
5:Simult Help
6:Quit PolySmlt

a2x^2+a1x+a0=0
a2 =1
a1 =-4
a0 =3
MAIN|DEGR|CLR|LOAD|SOLVE

a2x^2+a1x+a0=0
x1 =1
x2 =3
MAIN|COEFS|STOa|STOx|STOy

Examples from past IB papers

An easy one from May 2000 Paper 1

The diagram shows the graph of $y = x^2 - 2x - 8$. The graph crosses the x-axis at the point A, and has a vertex at B.

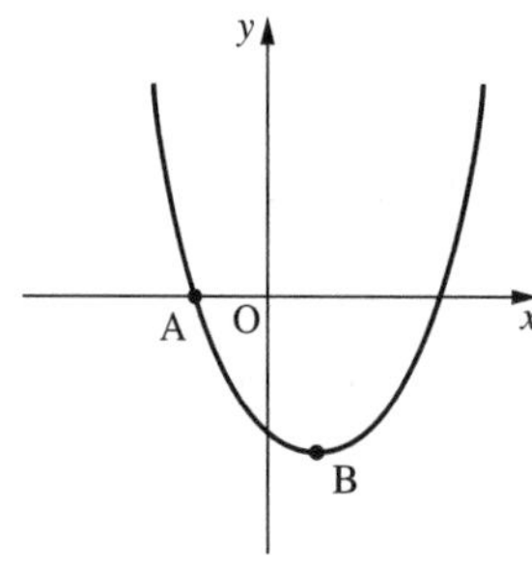

(a) Factorise $x^2 - 2x - 8$.

(b) Write down the coordinates of each of these points

(i) A;

(ii) B.

Working:

(a) $(x + 2)(x - 4)$

(b)(i) $x + 2 = 0$

$x = -2$

A is $(-2, 0)$

(ii) GDC: B is $(1, -9)$

Answers:

(a) $(x + 2)(x - 4)$

(b) (i) $(-2, 0)$

(ii) $(1, -9)$

A harder one from May 2005 Paper 1

A swimming pool is to be built in the shape of a letter L. The shape is formed from two squares with side dimensions x and $\sqrt{x}$ as shown.

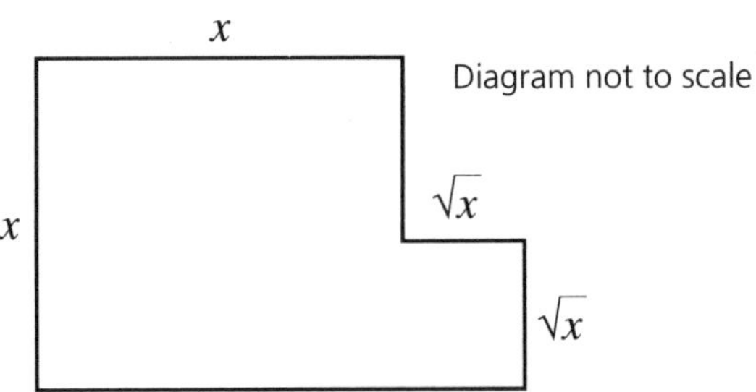

(a) Write down an expression for the area of the swimming pool surface.

(b) The area A is to be 30 m^2. Write a quadratic equation that expresses this information.

(c) Find both solutions of your equation in part (b).

(d) Which of the solutions in part (c) is the correct value of x for the pool? State briefly why you made this choice.

Working:

(a) $A = x^2 + x$

(b) $x^2 + x = 30$

(c) $x^2 + x - 30 = 0$

$(x + 6)(x - 5) = 0$

$x = -6$ or $x = 5$

(d) $x = 5$ because length must be positive

Answers:

(a) $A = x^2 + x$

(b) $x^2 + x = 30$

(c) $x = -6$ or $x = 5$

(d) $x = 5$ because length must be positive

Now you practise it

An easy one from November 1999 Paper 1

The perimeter of this rectangular field is 220 m. One side is x m as shown.

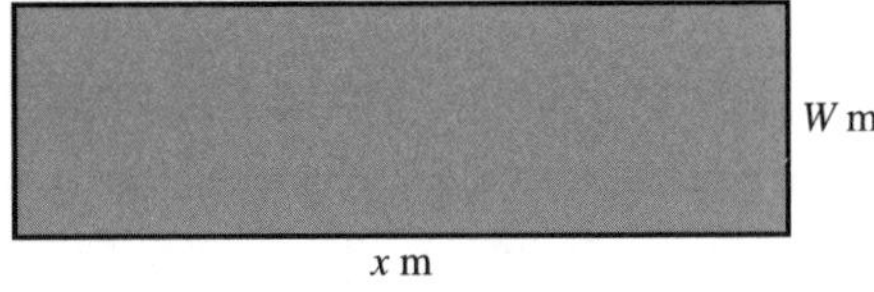

a) Express the width (W) in terms of x.

b) Write an expression, in terms of x only, for the area of the field.

c) If the length (x) is 70 m, find the area.

Working:

Answers:

(a) ....

(b) ....

(c) ....

A harder one from May 2002 Paper 1

a) Find the solutions of the equation $x^2 - 5x - 24 = 0$.

b) The equation $ax^2 - 9x - 30 = 0$ has solution $x = 5$ and $x = -2$. Find the value of a.

Working:

Answers:

(a) ....

(b) ....

Additional practice problems

1. From November 2006 Paper 2 *[Maximum mark: 9]*

The first three terms of an arithmetic sequence are

$$2k + 3,\ 5k - 2 \text{ and } 10k - 15.$$

(a) Show that $k = 4$. *[3 marks]*

(b) Find the values of the first three terms of the sequence. *[1 mark]*

(c) Write down the value of the common difference. *[1 mark]*

(d) Calculate the 20th term of the sequence. *[2 marks]*

(e) Find the sum of the first 15 terms of the sequence. *[2 marks]*

2. From November 1999 Paper 2 (modified) *[Maximum mark: 6]*

Angela's employer agrees to give her an interest free loan of $4000 to buy a car. The employer is to recover the money by making the following deductions from Angela's salary:

$*x* in the first month,
$*y* every subsequent month.

The total deductions after 20 months is $1540 and after 30 months is $2140.

(a) Find x and y. *[4 marks]*

(b) How many months will it take Angela to completely pay off the $4000 loan? *[2 marks]*

3. From May 2007 Paper 2 *[Maximum mark: 16]*

(i) The natural numbers: 1, 2, 3, 4, 5, ... form an arithmetic sequence.

(a) State the values of u_1 and d for this sequence. *[2 marks]*

(b) Use an appropriate formula to show that the sum of the natural numbers from 1 to n is given by $\frac{1}{2}n(n + 1)$. *[2 marks]*

(c) Calculate the sum of the natural numbers from 1 to 200. *[2 marks]*

(ii) A geometric progression G_1 has 1 as its first term and 3 as its common ratio.

(a) The sum of the first n terms of G_1 is 29 524. Find n. *[3 marks]*

A second geometric progression has the form $1, \frac{1}{3}, \frac{1}{9}, \frac{1}{27}, \ldots$

(b) State the common ratio for G_2. *[1 mark]*

(c) Calculate the sum of the first 10 terms of G_2. *[2 marks]*

(d) Explain why the sum of the first 1000 terms of G_2 will give the same answer as the sum of the first 10 terms, when corrected to three significant figures. *[1 mark]*

(e) Using your results from parts (a) to (c), or otherwise, calculate the sum of the first 10 terms of the sequence $2, 3\frac{1}{3}, 9\frac{1}{9}, 27\frac{1}{27}, \ldots$ Give your answer **correct to one decimal place.** *[3 marks]*

4. From May 2001 Paper 2 *[Maximum mark: 7]*

A picture is in the shape of a square of side 5 cm. It is surrounded by a wooden frame of width x cm, as shown in the diagram below.

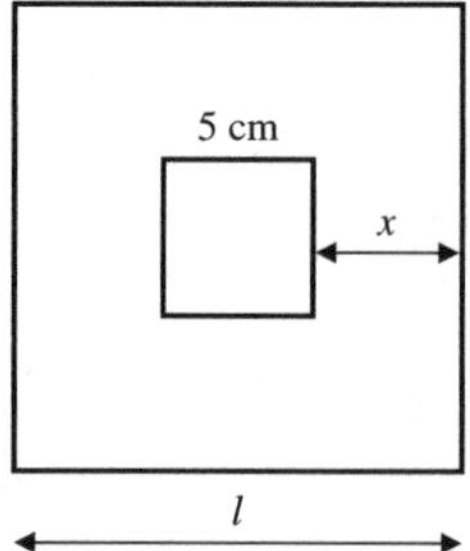

The length of the wooden frame is l cm, and the area of the wooden frame is A cm^2.

(a) Write an expression for the length l in terms of x. *[1 mark]*

(b) Write an expression for the area A in terms of x. *[2 marks]*

(c) If the area of the frame is 24 cm^2, find the value of x. *[4 marks]*

TOPIC 8

Set theory and Venn diagrams

What do you need to know?

You must memorise the basic set symbols:

$\cup$	union	to put two sets together without repeats
$\cap$	intersection	to find elements in common between two or more sets
$\in$	element	an element of a set
$\notin$	not an element	not an element of a set
$\subseteq$	subset	one set either inside or equal to another set
$\subset$	proper subset	one set totally inside another set
A'	complement	everything not in the set A
U	universal set	everything in all the sets

SCOTT SAYS:

A nice way to remember the symbols is to think of what they look like: $\cup$ is like a cup you put sets into, $\cap$ is the cup but you discard the extra stuff, $\in$ looks like an e for element, $\subset$ looks like a < sign (and means almost the same thing), and $\subseteq$ looks like $\leq$ (and again means almost the same thing).

More vocabulary – A *Venn diagram* is a diagram, usually of circles, that shows the relationship between two or more sets.

IRENE SAYS:

One thing I love to ask on IB exams is for students to draw Venn diagrams that do *not* look like three intersecting circles. Don't automatically draw those Olympic rings unless you are sure that is what's required!

How do you do it?

Usually the exam will ask you to either write out the elements of a set, draw or interpret a Venn diagram, or write the set symbols.

Examples

(The shaded regions are the answers.)

If U = {positive integers less than 10}, A = {2, 3, 5, 7} and B = {2, 4, 6, 8}, then

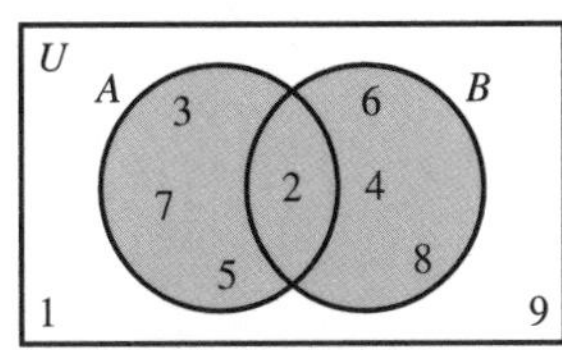

$A \cup B = \{2,3,4,5,6,7,8\}$

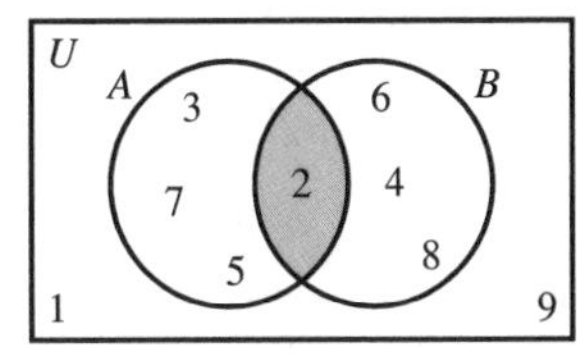

$A \cap B = \{2\}$

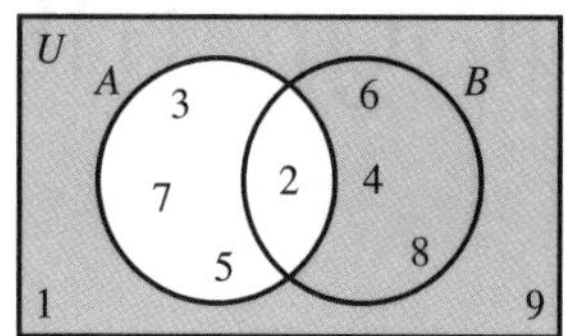

$A' = \{1,4,6,8,9\}$

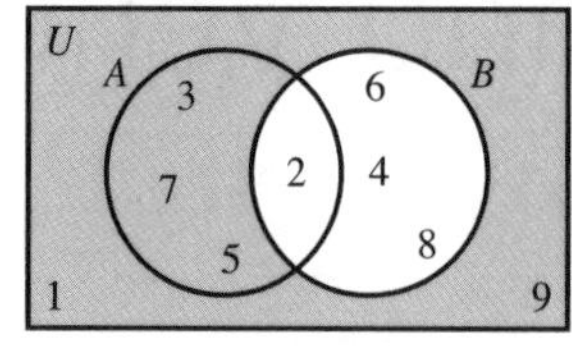

$B' = \{1,3,5,7,9\}$

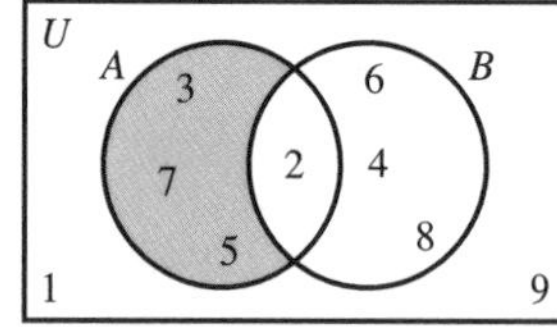

$A \cap B' = \{3,5,7\}$

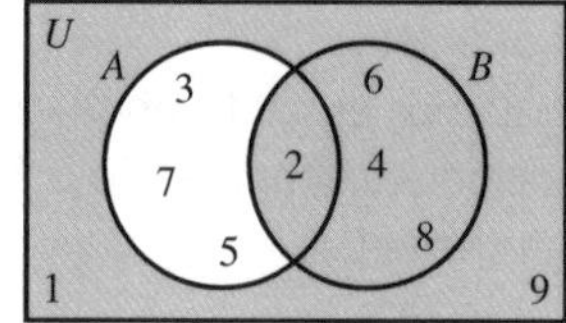

$A' \cup B = \{1,2,4,6,8,9\}$

HANA AND CHRIS SAY:

A good way to do these problems is to start "from the inside out". Write down what all the sets have in common, then do the rest of the intersections, then the remainder of the sets and finally what is left in the universal set. Remember: *go from the inside out!*

How can my calculator help me do this?

Your calculator is useless for this topic!

Examples from past IB papers

An easy one from May 2003 Paper 1

In the Venn diagram below, *A*, *B* and *C* are subsets of a universal set $U = \{1, 2, 3, 4, 6, 7, 8, 9\}$

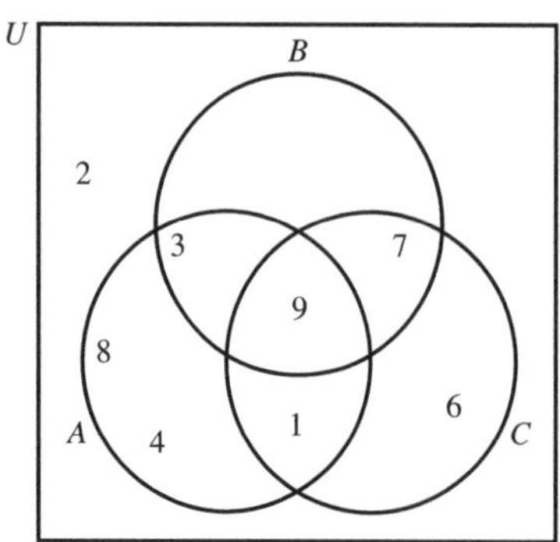

List the elements in each of the following sets.

(a) $A \cup B$

(b) $A \cap B \cap C$

(c) $(A' \cap C) \cup B$

Working:

(c) $A' = \{2, 6, 7\}$

$A' \cap C = \{6, 7\}$

$(A' \cap C) \cup B = \{3, 6, 7, 9\}$

Answers:

(a) $\{1, 3, 4, 7, 8, 9\}$

(b) $\{9\}$

(c) $\{3, 6, 7, 9\}$

A harder one from November 2002 Paper 1

A poll was taken of the leisure time activities of 90 students.

60 students watch TV (T), 60 students read (R), 70 students go to the cinema (C).
26 students watch TV, read **and** go to the cinema.
20 students watch TV and go to the cinema only.
18 students read and go to the cinema only.
10 students read and watch TV only.

(a) Draw a Venn diagram to illustrate the above information.

(b) Calculate how many students

(i) only watch TV;

(ii) only go to the cinema.

Working:

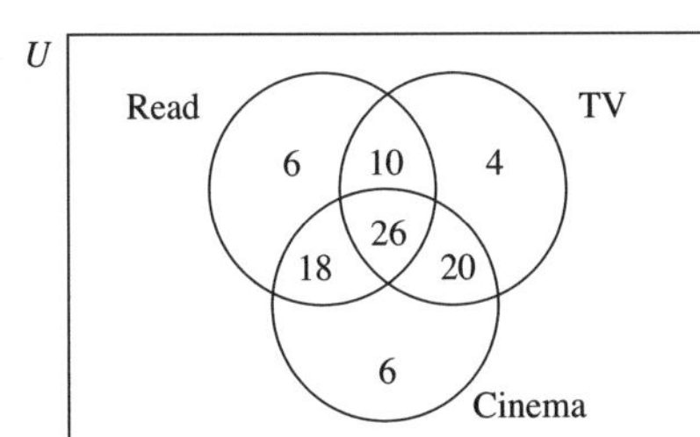

Working:

(b) (i) $60 - (10 + 26 + 20)$
$= 4$

(ii) $70 - (18 + 26 + 20)$
$= 6$

Answers:

(a) See diagram

(b) (i) 4

(ii) 6

Now you practise it

An easy one from November 2004 Paper 1

Shade the given region on the corresponding Venn diagram.

(a) $A \cap B$

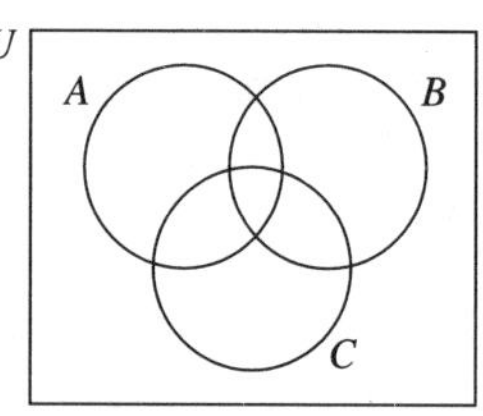

(b) $C \cup B$

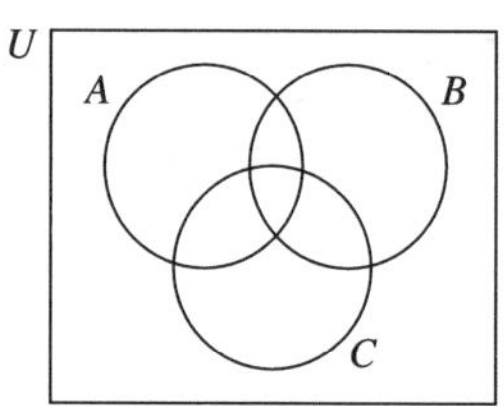

(c) $(A \cup B \cup C)'$

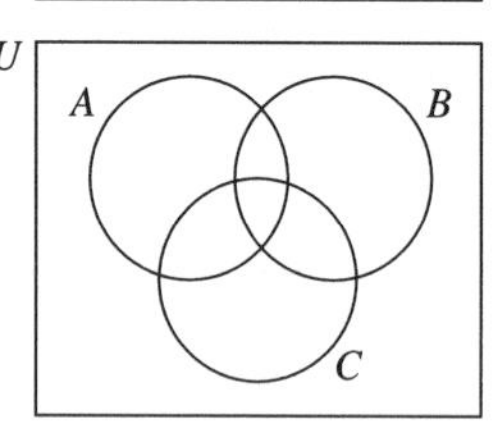

(d) $A \cap C'$

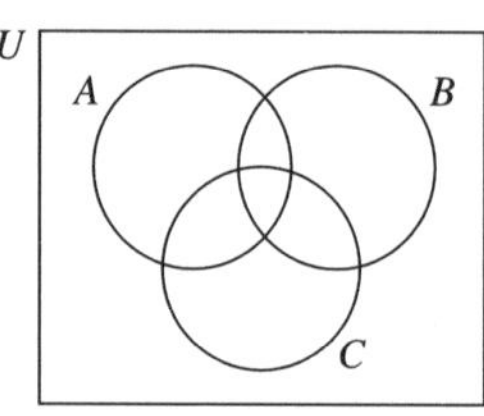

A harder one from November 2007 Paper 1

The universal set U is the set of integers from 1 to 20 inclusive.

A and B are subsets of U where:
A is the set of even numbers between 7 and 17.
B is the set of multiples of 3.

List the elements of the following sets:

(a) A

(b) B

(c) $A \cup B$

(d) $A \cap B'$

Working:

Answers:

(a) ....

(b) ....

(c) ....

(d) ....

TOPIC 9

Logic symbols and statements

What do you need to know?

You must memorise the basic logic symbols:

$\neg$	negation	"not . . ." or "it is not true that . . ."
$\wedge$	conjunction	". . . and . . ."
$\vee$	disjunction	" . . . or . . ."
$\veebar$	exclusive disjunction	" . . . or . . . but not both"
$\Rightarrow$	implication	"If . . . then . . ."
$\Leftrightarrow$	equivalence	" . . . if and only if . . ."

and the definitions of implication terms:

converse	when you switch the position of the two statements or switch the arrow direction ($p \Rightarrow q$ becomes $q \Rightarrow p$ or $p \Leftarrow q$)
inverse	when you negate both statements ($p \Rightarrow q$ becomes $\neg p \Rightarrow \neg q$)
contrapositive	when you switch the position of the two statements **and** negate both statements ($p \Rightarrow q$ becomes $\neg q \Rightarrow \neg p$)

Very important: **the converse is logically equivalent to the inverse,** and **the contrapositive is logically equivalent to the original implication!!**

In symbols, that means $q \Rightarrow p$ is the same as $\neg p \Rightarrow \neg q$, and $p \Rightarrow q$ is the same as $\neg q \Rightarrow \neg p$.

How do you do it?

Usually the exam will ask you to either write out the words from a given statement in symbols or write the symbols from a given statement in words. Sometimes it will ask you to write the converse, inverse or contrapositive of a statement.

Examples

If p represents the statement, "I will pass this Math Studies exam.", q represents, "My mum will buy me a new car." and r represents, "I am going on a trip to France.", then

1 "It is not true that I will pass this Math Studies exam or my mum will buy me a new car, but not both" translates into symbols as $\neg(p \veebar q)$.

2 $\neg q \Leftrightarrow \neg p$ translates into words as, "My mum will not buy me a new car if and only if I do not pass this Math Studies exam".

3 "If I do not pass this Math Studies exam, then either I am not going on a trip to France or my mum will not buy me a new car" translates into symbols as $\neg p \Rightarrow (\neg r \vee \neg q)$.

4 The converse of "If my mum buys me a new car then I passed the Math Studies exam" becomes "If I passed the Math Studies exam then my mum will buy me a new car."

SCOTT SAYS:

Don't get creative with your words. When you write these statements, copy the words *exactly* as you see them. Don't paraphrase!!

IRENE SAYS:

Warning! Many students try to "distribute" a negation sign over brackets. It doesn't work the way you think! $\neg(p \wedge q)$ is not equal to $\neg p \wedge \neg q$. It's actually equal to $\neg p \vee \neg q$ by a rule called "De Morgan's Law" which is not on the syllabus. Just be careful!

How can my calculator help me do this?

Your calculator is useless for this topic!

Examples from past IB papers

An easy one from May 2004 Paper 1

Let p and q be the statements:

p: *Sarah eats lots of carrots.*
q: *Sarah can see well in the dark.*

Write the following statements in words:

(a) $p \Rightarrow q$

(b) $\neg p \wedge q$

(c) Write the following statement in symbolic form.

If Sarah cannot see well in the dark, then she does not eat lots of carrots.

(d) Is the statement in part (c) the *inverse*, the *converse* or the *contrapositive* of the statement in part (a)?

Working:

Answers:

(a) *If Sarah eats lots of carrots, then she can see well in the dark.*

(b) *Sarah does not eat lots of carrots and she can see well in the dark.*

(c) $\neg q \Rightarrow \neg p$

(d) *contrapositive*

IRENE SAYS:

You probably noticed that I did not give you a lot of space to write your answer in (b). Don't worry – just continue writing outside the lines or just write the answer below and put an arrow to where the answer should be.

A harder one from November 1999 Paper 1

Three propositions p, q and r are defined as follows:

p: the water is cold q: the water is boiling r: the water is warm

(a) Write one sentence in words for the following logic statement.

$$(\neg p \wedge \neg q) \Rightarrow r$$

(b) Write the following sentence as a logic statement using symbols only.

"The water is cold if and only if it is neither boiling nor warm."

Working:

Answers:

(a) *If the water is not cold and not boiling then it is warm*

(b) $p \Leftrightarrow \neg q \wedge \neg r$

Now you practise it

An easy one from May 2002 Paper 1

Consider the statement *"If a figure is a square, then it is a rhombus."*

(a) For this statement, write in words

(i) its converse;

(ii) its inverse;

(iii) its contrapositive.

(b) Only one of the statements in part (a) is true. Which one is it?

Working:

Answers:

(a) (i) ....

(ii) ....

(iii) ....

(b) ....

A harder one from November 2006 Paper 1

Two logic propositions are given

p: *Dany goes to the cinema.*
q: *Dany studies for the test.*

(a) Write in words the proposition

$$p \veebar q$$

(b) Given the statement *s*: *"If Dany goes to the cinema then Dany doesn't study for the test."*

(i) Write *s* in symbolic form.

(ii) Write in symbolic form the contrapositive of part (b)(i).

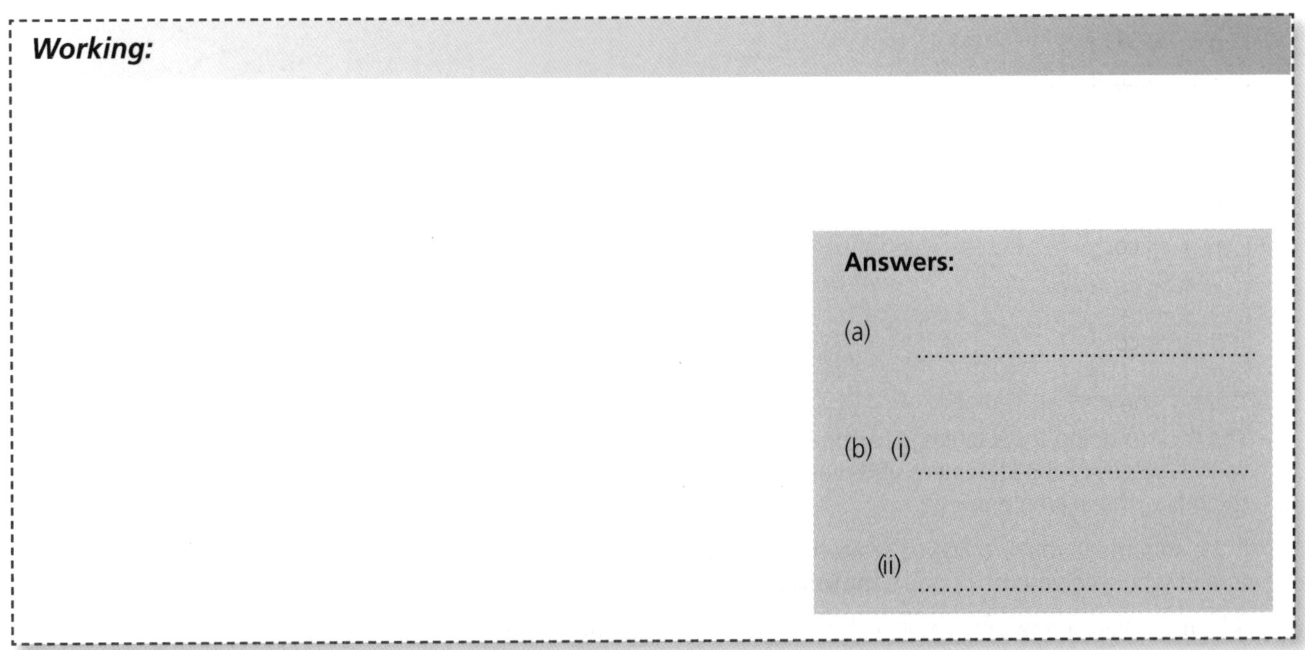

Truth tables

What do you need to know?

You must know the basic truth tables:

p	q	$\neg p$	$p \wedge q$	$p \vee q$	$p \veebar q$	$p \Rightarrow q$	$p \Leftrightarrow q$
T	T	F	T	T	F	T	T
T	F	F	F	T	T	F	F
F	T	T	F	T	T	T	F
F	F	T	F	F	F	T	T

SCOTT SAYS:

This is how to remember $p \Rightarrow q$:

Imagine a husband says to his wife, "If I get a raise at work, I will buy you a diamond ring." The question is: when is the husband good or bad?

Let p = I get a raise and q = I buy you a ring. If he gets the raise and buys the ring, he's a good guy (p is true, q is true, thus $p \Rightarrow q$ is true).

If he gets the raise but does not buy the ring, he's a bad guy (p is true, q is false, thus $p \Rightarrow q$ is false).

If he does not get the raise but buys the ring anyway, he's definitely a good guy (p is false, q is true, thus $p \Rightarrow q$ is true). And lastly if he does not get the raise and does not buy the ring, he's still a good guy (p is false, q is false, thus $p \Rightarrow q$ is true).

Try it – it works!

For truth tables with three variables, you set them up like this:

p	q	r
T	T	T
T	T	F
T	F	T
T	F	F
F	T	T
F	T	F
F	F	T
F	F	F

The trick to doing truth tables is to "build" them. Start by making columns of the various pieces of the statement and then put them together to create the whole thing (see the example below).

If the statement you're trying to prove has a truth table where all the values are true (a whole column of T), it is a **tautology**.

If all the values are false (a whole column of F), it is a **contradiction**.

If two columns of a truth table are identical, we say that the two statements at the top are **logically equivalent**.

If you want to prove that an argument is valid, it's an if-then statement where all the pieces of the argument are "AND"ed together on the "if" side and the answer is by itself on the "then" side:

[{statement 1} $\wedge$ {statement 2} $\wedge$ {statement 3} $\wedge$ . . .] $\Rightarrow$ {what you want to prove}
Example

Here is a truth table for $(p \wedge \Rightarrow q) \Leftrightarrow (q \veebar p)$:

p	q	$\Rightarrow q$	$p \wedge \Rightarrow q$	$q \veebar p$	$(p \wedge \Rightarrow q) \Leftrightarrow (q \veebar p)$
T	T	F	F	F	T
T	F	T	T	T	T
F	T	F	F	T	F
F	F	T	F	F	T

Prove the argument valid or invalid: If you are human, then you are a mammal. If you are not a mammal, then you do not have live births. You have live births. Therefore you are human.

Let p = "You are human." Let q = "You are a mammal." Let r = "You have live births."

The truth table that we need to set up is: $\{(p \Rightarrow q) \wedge (\neg q \Rightarrow \neg r) \wedge r\} \Rightarrow p$

p	q	r	$\neg q$	$\neg r$	$p \Rightarrow q$	$\neg q \Rightarrow \neg r$	$(p \Rightarrow q) \wedge (\neg q \Rightarrow \neg r) \wedge r$	$\{(p \Rightarrow q) \wedge (\neg q \Rightarrow \neg r) \wedge r\} \Rightarrow p$
T	T	T	F	F	T	T	T	T
T	T	F	F	T	T	T	F	T
T	F	T	T	F	F	F	F	T
T	F	F	T	T	F	T	F	T
F	T	T	F	F	T	T	T	F
F	T	F	F	T	T	T	F	T
F	F	T	T	F	T	F	F	T
F	F	F	T	T	T	T	F	T

For the argument to be valid, it needs to be a tautology. So because of that *one* F you see in the last column, this argument is **not** valid.

SCOTT SAYS:

If you get confused trying to work with big truth tables, try marking the tops of the columns as you need them.

For example, if you are doing the last column, you draw an arrow from the second-to-last column to the first column. This helps you see what you are doing. Now just remember the story of "the raise and the ring" as you follow the arrow and you'll never make a mistake!

How can my calculator help me do this?

Your calculator is pretty useless for this topic! There is a way to do truth tables on your GDC but it is not worth the effort, believe me.

Examples from past IB papers

An easy one from May 2001 Paper 1

$[(p \Leftrightarrow q) \wedge p] \Rightarrow q$

(a) Complete the truth table below for the compound statement above.

p	q	$p \Leftrightarrow q$	$(p \Leftrightarrow q) \wedge p$	$[(p \Leftrightarrow q) \wedge p] \Rightarrow q$
T	T			
T	F			
F	T			
F	F			

(b) Explain the significance of your result.

Working:

p	q	$p \Leftrightarrow q$	$(p \Leftrightarrow q) \wedge p$	$[(p \Leftrightarrow q) \wedge p] \Rightarrow q$
T	T	T	T	T
T	F	F	F	T
F	T	F	F	T
F	F	T	F	T

Answers:

(b) *It's a tautology*

A harder one from May 2007 Paper 1

The truth table below shows the truth values for the proposition

p	q	$\neg p$	$\neg q$	$p \veebar q$	$\neg p \veebar \neg q$	$p \veebar q \Rightarrow \neg p \veebar \neg q$
T	T	F	F		F	
T	F	F		T	T	T
F	T	T	F	T	T	T
F	F	T	T	F		T

(a) Explain the distinction between the compound propositions, $p \veebar q$ and $p \vee q$.

(b) Fill in the four missing truth-values on the table.

(c) State whether the proposition $p \veebar q \Rightarrow \neg p \veebar \neg q$ is a tautology, a contradiction or neither.

Working:

$\neg q$	$p \veebar q$	$p \veebar q \neg p$	$p \veebar q \Rightarrow \neg p \veebar \neg q$
F	F	F	T
T	T	T	T
F	T	T	T
T	F	F	T

Answers:

(a) Both mean "p or q" but the first means not both

(c) tautology

HANA AND CHRIS SAY:

Did you notice here that the question has a typo? There should be parenthesis/brackets around $p \veebar q$ and around $\neg p \veebar \neg q$. It should look like this: $(p \veebar q) \Rightarrow (\neg p \veebar \neg q)$. Oops! Believe it or not, *it is very possible that you will see a typo on your exam.* Don't panic! Just answer it as best as you can and make sure to tell your teacher about it afterwards so she can complain to the IB!

Now you practise it

An easy one from May 2006 Paper 1

Consider the statements

p : *The sun is shining.*
q : *I am wearing my hat.*

(a) Write down, in words, the meaning of $q \Rightarrow \neg p$.

(b) Complete the truth table.

p	q	$\neg p$	$q \Rightarrow \neg p$
T	T		
T	F		
F	T		
F	F		

(c) Write down, in symbols, the converse of $q \Rightarrow \neg p$.

Working:

Answers:

(a) ....

(c) ....

A harder one from November 2005 Paper 1

Complete the truth table for the compound proposition $(p \wedge \neg q) \Rightarrow (p \vee q)$.

p	q	$\neg q$	$(p \wedge \neg q)$	$(p \vee q)$	$(p \wedge \neg q) \Rightarrow (p \vee q)$
T	T	F	F		
T	F	T	T		
F	T	F		T	
F	F		F	F	

Working:

TOPIC 11

Probability: simple problems

What do you need to know?

Simple probability problems usually only involve one principle: the probability of an event is the ratio of the number of successful events to the total number of events possible:

$$\text{Probability of an event E} = \frac{\text{number of successful events}}{\text{total number of events possible}}$$

You can write your probabilities as a fraction, a decimal or a percentage. The IB accepts all three forms. I recommend you use fractions if at all possible as it will make your life easier when doing more complex problems.

IRENE SAYS:

Don't worry about simplifying fractions when doing probability on the IB exam. I don't take off marks if you leave your fractions unsimplified! Better to leave it this way than to simplify incorrectly!

The probability of an event ranges from a minimum of 0 (can't happen – an *impossible* event) to 1 (must happen – a *certain* event). *All probabilities* must be between 0 and 1 inclusive!

If you want to find the probability of something not happening, just subtract the probability of it happening from 1:

Probability of an event E *not* happening = 1 – probability of E happening

How do you do it?

IB exam questions involving simple probability will give you a table, a Venn diagram or another way to show the relevant information. You just have to count the number of successful events and the total number of possible events, and then use the formula above.

Sometimes it is useful to make a *table of outcomes* of the problem – a list of all the possibilities. For example, here is a sample space of what can happen if you roll two dice:

1,1	1,2	1,3	1,4	1,5	1,6
2,1	2,2	2,3	2,4	2,5	2,6
3,1	3,2	3,3	3,4	3,5	3,6
4,1	4,2	4,3	4,4	4,5	4,6
5,1	5,2	5,3	5,4	5,5	5,6
6,1	6,2	6,3	6,4	6,5	6,6

Clearly there are 36 possible events.

Examples

1. The table shows the participation in sports of some students:

	Basketball	Soccer
Girls	18	23
Boys	27	22

Say you pick a student at random.

The probability of picking a girl who plays basketball is 18 out of 90, or $\frac{18}{90}$.

The probability of picking a soccer player is $\frac{45}{90}$ (or $\frac{1}{2}$ or 0.5 or 50%).

The probability of picking a student who is not a boy basketball player is $1 - \frac{27}{90} = \frac{63}{90}$.

2. Given the sample space above for rolling two dice, the probability of rolling a double is $\frac{6}{36}$ (or $\frac{1}{6}$ or 0.167 or 16.7%).

   The probability of rolling a 5 on one of the dice is $\frac{11}{36}$ (make sure you do not count the 5,5 possibility twice!).

   The probability of not rolling a prime number on either die is $1 - \frac{27}{36} = \frac{9}{36}$.

How can my calculator help me do this?

The only way the calculator is helpful here is for working with fractions. You can use the Math → Frac feature.

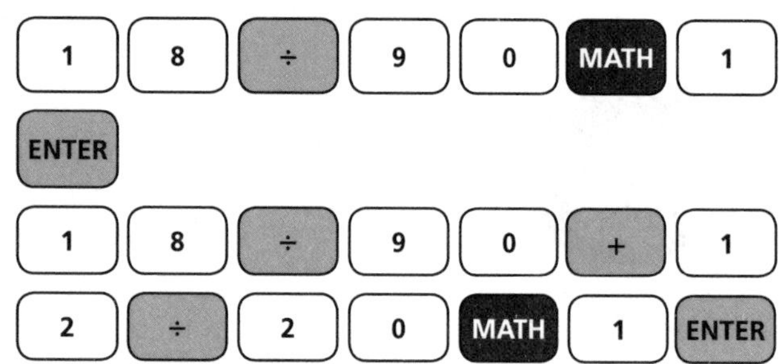

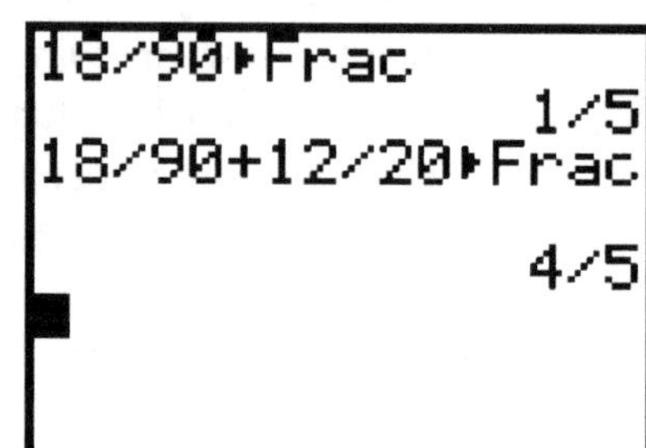

Examples from past IB papers

An easy one from November 2002 Paper 1

Heinrik rolls two 6-sided dice at the same time. One die has three red sides and three black sides. The other die has the sides numbered from 1 to 6. By means of a tree diagram, table of outcomes or otherwise, answer each of the following questions.

(a) How many different possible combinations can he roll?

(b) What is the probability that he will roll a red and an even number?

(c) What is the probability that he will roll a red or black and a 5?

(d) What is the probability that he will roll a number less than 3?

Working:

(a) R1, R2, R3, R4, R5, R6, B1, B2, B3, B4, B5, B6

(b) R2, R4, R6: $\frac{3}{12} = \frac{1}{4}$

(c) R5, B5: $\frac{2}{12} = \frac{1}{6}$

(d) R1, R2, B1, B2: $\frac{4}{12} = \frac{1}{3}$

Answers:

(a) 12

(b) $\frac{1}{4}$

(c) $\frac{1}{6}$

(d) $\frac{1}{3}$

A harder one from May 2000 Paper 1

On a certain game show, contestants spin a wheel to win a prize, as shown in the diagram. The larger angles are **40°** (the shaded sectors), and the smaller angles are **20°**.

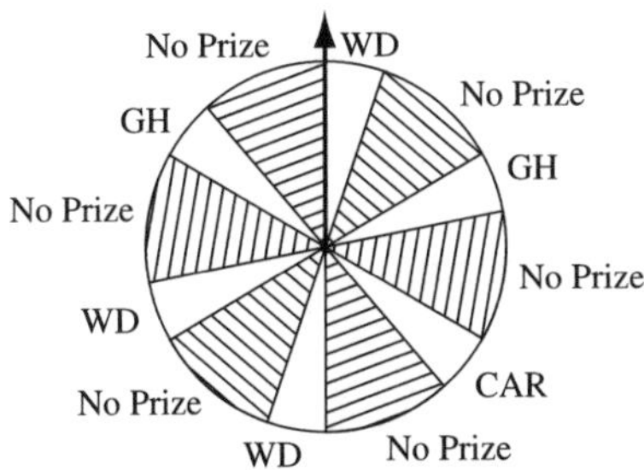

Find the probability that a contestant

(a) will **not** win a prize;

(b) will win a holiday in Greece (GH);

(c) will win a washer/dryer (WD), given that he knows that he has won a prize;

(d) will win a holiday in Greece **or** a washer/dryer.

Working:

(a) $\frac{(40 \times 6)}{360} = \frac{240}{360} = \frac{2}{3}$

(b) $\frac{(20 \times 2)}{360} = \frac{40}{360} = \frac{1}{9}$

(c) $\frac{(20 \times 3)}{(20 \times 6)} = \frac{60}{120} = \frac{1}{2}$

(d) $\frac{(20 \times 5)}{360} = \frac{100}{360} = \frac{5}{18}$

Answers:

(a) $\frac{2}{3}$

(b) $\frac{1}{9}$

(c) $\frac{1}{2}$

(d) $\frac{5}{18}$

SCOTT SAYS:

Part (c) is tricky here – it involves "conditional" probability that is actually covered in depth in the next section. But give it a try . . . it's not too hard if you just think about it for a minute! Hint: the denominator is no longer 360°.

Now you practise it

An easy one from May 1996 Paper 1

A basket of fruit contains 10 apples, 6 bananas and 4 oranges. A fruit is selected at random.

(a) Find the probability that the first fruit selected is not an orange.

The first and the second fruits selected are both bananas. They are eaten.

(b) Find the probability that the next fruit selected will be an apple.

Working:

Answers:

(a) ....

(b) ....

A harder one from May 2000 Paper 1

Of a group of five students, two will be selected to visit the United Nations. The five students are John, Maria, Raul, Henri and Susan.

(a) With the aid of a tree diagram or a table of outcomes, find the number of **different** possible combinations of students that could go to the United Nations.

(b) Find the probability that both Maria and Susan will go on the trip.

Working:

Answers:

(a) ....

(b) ....

TOPIC 12

Probability: harder problems (compound and conditional probability)

What do you need to know?

Compound probability problems usually require you to **multiply** two or more probabilities together.

Conditional probability problems usually require you to **divide** two probabilities.

You need to remember that if a problem asks you to find the probability of two or more things happening (e.g. rolling two dice, raining and eating an ice cream cone, etc . . .), it is a compound probability problem. If a problem limits your sample space, usually with the words "given that", it is a conditional probability problem. Once you know what kind of problem it is, it is straightforward to solve.

How do you do it?

The hard part of probability is that there are lots of ways to solve the same problem. It all depends on how you *see* the problem. The method that works the best in most situations is to *draw a tree diagram*. If you are working out probabilities in a tree, multiply the probabilities across the branches to the right. Once you have done that, add all the products that are relevant to the problem going down.

When doing conditional probability, the new denominator is the sum of the probabilities of the condition (whatever follows "given that" or something similar). It's easier to see with an example:

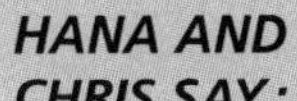

Remember these two mantras: MULTIPLY ACROSS AND ADD DOWN (for compound) and CHANGE THE DENOMINATOR (for conditional). Say them to yourself every time you draw a tree diagram and you will never go wrong.

Examples

The probability of rain tomorrow is 15%. If it rains tomorrow, a person has a 72% probability of being late for work. If it does not rain tomorrow, a person has an 18% probability of being late for work.

Tree diagram:

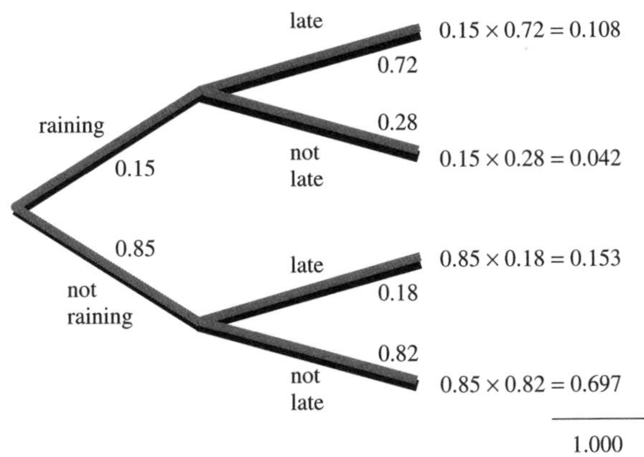

The probability of rain and being late for work is $0.15 \times 0.72 = 0.108$ or 10.8%.

The probability of being late is $(0.15 \times 0.72) + (0.85 \times 0.18) = 0.108 + 0.153 = 0.261$ or 26.1%.

The probability of rain given that the person is late for work is $0.108 \div (0.108 + 0.153) = 0.414$ (3sf) or 41.4%.

SCOTT SAYS:

Notice that the probabilities in each set of branches must add up to 1, and the total sum going down the right-hand side also adds up to 1. Of course! If you take into account every possibility, all the probabilities must add up to 1.

How can my calculator help me do this?

Unfortunately your calculator is not very helpful here except for doing arithmetic and for dealing with fractions. Sorry!

Examples from past IB papers

An easy one from November 2001 Paper 1

A teacher has a box containing six type A calculators and four type B calculators.

The probability that a type A calculator is faulty is 0.1 and the probability that a type B calculator is faulty is 0.12.

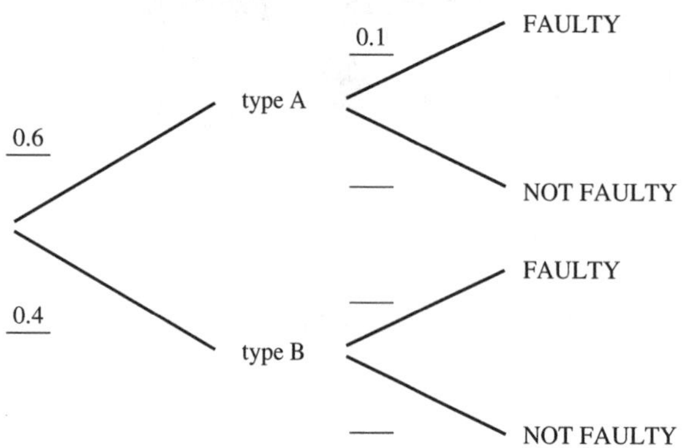

(a) Complete the tree diagram, showing all the probabilities.

(b) A calculator is selected at random from the box. Find the probability that the calculator is

(i) a faulty type A;

(ii) not faulty.

Working:

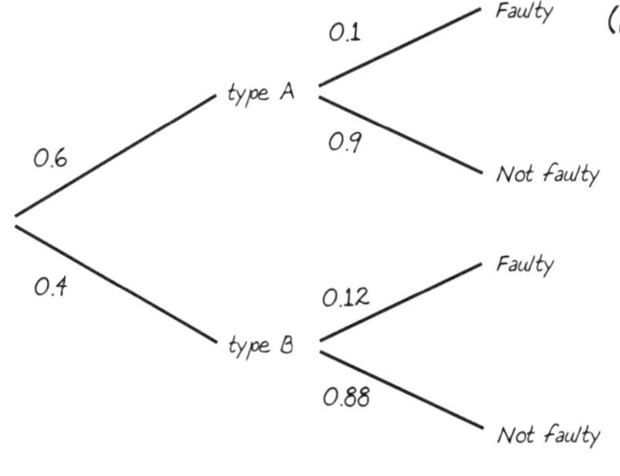

(b) (i) $0.6 \times 0.1 = 0.06$

(ii) $0.6 \times 0.9 + 0.4 \times 0.88 = 0.892$

Answers:

(b) (i) 0.06

(ii) 0.892

A harder one from May 1999 Paper 1

The table below shows the number of left and right handed tennis players in a sample of 50 males and females.

	Left handed	Right handed	Total
Male	3	29	32
Female	2	16	18
Total	5	45	50

If a tennis player was selected at random from the group, find the probability that the player is

(a) male and left handed;

(b) right handed;

(c) right handed, given that the player selected is female.

Working:

Answers:

(a) $\frac{3}{50}$

(b) $\frac{45}{50}$ or $\frac{9}{10}$

(c) $\frac{16}{18}$ or $\frac{8}{9}$

Now you practise it

An easy one from November 2005 Paper 1

Jim drives to work each day through two sets of traffic lights.

The probability of the first set of traffic lights being red is 0.65.

If the first set is red then the probability that the next set of traffic lights is red is 0.46.

If the first set is not red, the probability that the next set is red is 0.72.

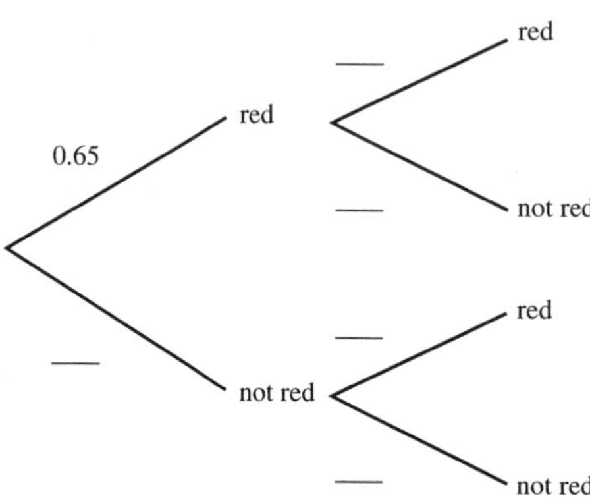

(a) Complete the tree diagram above.

(b) Calculate the probability that the second set of traffic lights is red.

Working:

Answers:

(b) ....

A harder one from November 1999 Paper 1

A bag contains 2 red, 3 yellow and 5 green sweets.

Without looking, Mary takes one sweet out of the bag and eats it. She then takes out a second sweet.

(a) If the first sweet is green, what is the probability that the second sweet is also green?

(b) If the first sweet is not red, what is the probability that the second sweet is red?

Working:

Answers:

(a) ....

(b) ....

Additional practice problems

1. From May 1999 Paper 2 *[Maximum mark: 20]*

(i) A box contains 10 coloured light bulbs, 5 green, 3 red and 2 yellow. One light bulb is selected at random and put into the light fitting of room A.

(a) What is the probability that the light bulb selected is

(i) green? *[1 mark]*

(ii) not green? *[1 mark]*

A second light bulb is selected at random and put into the light fitting in room B.

(b) What is the probability that

(i) the second light bulb is green given the first light bulb was green? *[1 mark]*

(ii) both light bulbs were not green? *[2 marks]*

(iii) one room had a green light bulb and the other room does not have a green light bulb? *[3 marks]*

A third light bulb is selected at random and put in the light fitting of room C.

(c) What is the probability that

(i) all three rooms have green light bulbs? *[2 marks]*

(ii) only one room has a green light bulb? *[3 marks]*

(iii) at least one room has a green light bulb? *[2 marks]*

(ii) It is known that 5% of all AA batteries made by Power Manufacturers are defective. AA batteries are sold in packs of 4.

Find the probability that a pack of 4 has

(a) exactly two defective batteries; *[3 marks]*

(b) at least one defective battery. *[2 marks]*

2. From November 2001 Paper 2 *[Maximum mark: 13]*

(i) 100 students were asked which television channel (MTV, CNN or BBC) they had watched the previous evening. The results are shown in the Venn diagram below.

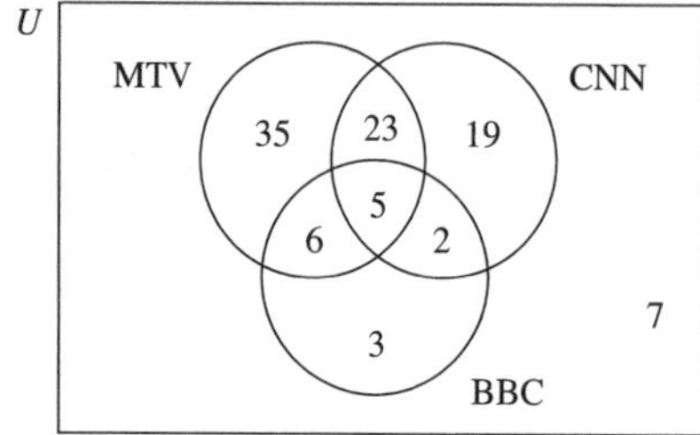

From the information in the Venn diagram, write down the number of students who watched

(a) both MTV and BBC;

(b) MTV or BBC;

(c) CNN and BBC but not MTV;

(d) MTV or CNN but not BBC. *[4 marks]*

(ii) Let p and q be the statements

p: you watch the music TV channel
q: you like music

(a) Consider the following logic statement.

If you watch the music TV channel then you like music.

(i) Write down in words the inverse of the statement.

(ii) Write down in words the converse of the statement. *[4 marks]*

(b) Construct truth tables for the following statements:

(i) $p \Rightarrow q$

(ii) $\neg p \Rightarrow \neg q$

(iii) $p \vee \neg q$

(iv) $\neg p \wedge q$ *[4 marks]*

(c) Which of the statements in part (b) are logically equivalent? *[1 mark]*

3. From May 2002 Paper 2 *[Maximum mark: 15]*

The Venn diagram below shows the number of students studying Science (S), Mathematics (M) and History (H) out of a group of 20 college students. Some of the students do not study any of these subjects, 8 study Science, 10 study Mathematics and 9 study History.

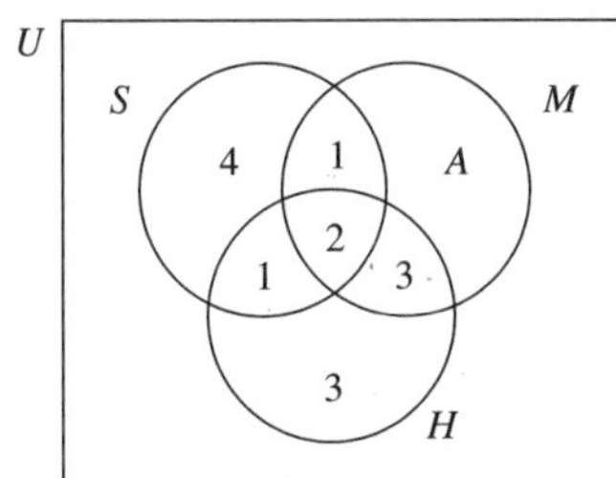

(a) (i) How many students belong to the region labelled A?

(ii) Describe in words the region labelled A.

(iii) How many students do not study any of the three subjects? *[5 marks]*

(b) Draw a sketch of the Venn diagram above and shade the region which represents $S' \cap H$. *[1 mark]*

(c) Calculate $n(S \cup H)$. *[2 marks]*

This group of students is to compete in an annual quiz evening which tests knowledge of Mathematics, Science and History. The names of the twenty students are written on pieces of paper and then put into a bag.

(d) One name is randomly selected from the bag. Calculate the probability that the student selected studies

(i) all three subjects;

(ii) History or Science. *[2 marks]*

(e) A team of two students is to be randomly selected to compete in the quiz evening. The first student selected will be the captain of the team. Calculate the probability that

(i) the captain studies all three subjects and the other team member does not study any of the three subjects;

(ii) one student studies Science only and the other student studies History only;

(iii) the second student selected studies History, given that the captain studies History and Mathematics. *[5 marks]*

Definition of a function, mapping diagrams, domain and range

What do you need to know?

A function is a kind of *mathematical machine*: you put in a number and a number comes out. Sometimes the number that comes out is the same as the number you put in, sometimes it is different. Each function machine has a *rule* – the rule tells the function what to do with the number.

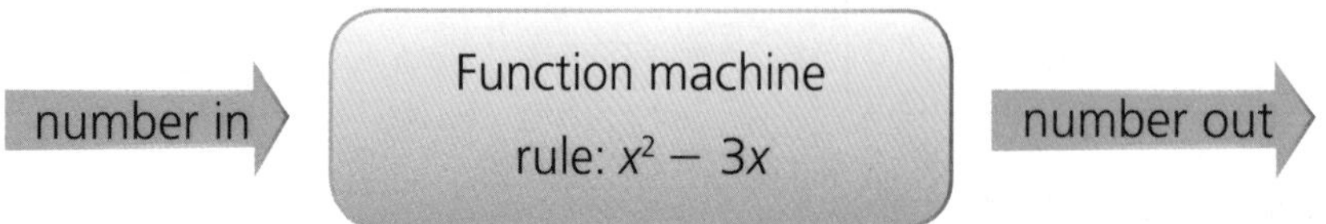

So if you put in a 1, you would get out $1^2 - 3 \cdot 1 = -2$.

We write this in several ways:

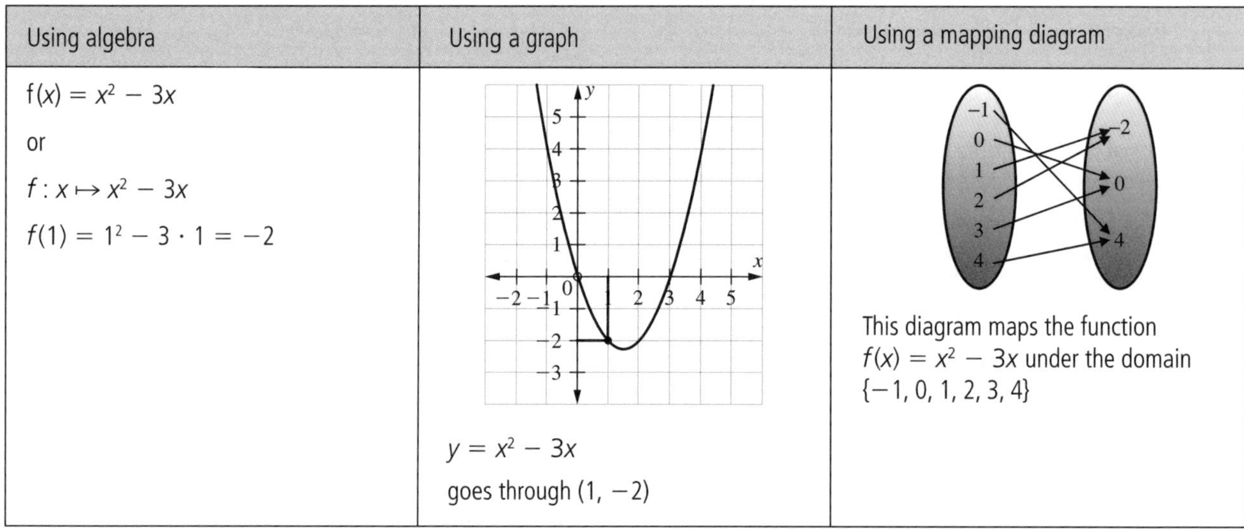

Using algebra	Using a graph	Using a mapping diagram
$f(x) = x^2 - 3x$ or $f : x \mapsto x^2 - 3x$ $f(1) = 1^2 - 3 \cdot 1 = -2$	$y = x^2 - 3x$ goes through $(1, -2)$	This diagram maps the function $f(x) = x^2 - 3x$ under the domain $\{-1, 0, 1, 2, 3, 4\}$

The *domain* of a function is the set of all numbers that can go **in** the function; the *range* of a function is the set of all numbers that can come **out** of the function.

Note: A function can have only one number that can come out for any number that goes in. In other words, suppose you put a number (like 1) into a function and get a number out (like −2). If you put in 1 again, you should get −2 out again! This is basically the definition of a function and you can see this by doing a *vertical line test* on a graph: if you can draw a vertical line anywhere on a graph and this line only touches the curve once, it is a function.

How do you do it?

IB questions want you to be familiar with the notation, to be able to put numbers into a function and to find the domain and range.

- To evaluate a function at a point, take that number and substitute it for x everywhere you see it in the function.
- To do the opposite (find x), set the function equal to the number given and solve for x.
- To find the domain of a function, you have to either think about all the numbers that could possibly go into the function (very hard to do), look at the x-axis on

a graph and see where the graph is covered (much easier, especially when you do this on your GDC), or list all the numbers on the left-hand side of a mapping diagram.

- To find the range of a function, you have to either think about all the numbers that could possibly come out of the function (again, very hard to do), look at the y-axis on a graph and see where the graph is covered (again easy with a GDC), or list all the numbers on the right-hand side of a mapping diagram.

Example

Given the function $f(x) = \sqrt{x+1}$, $f(8) = \sqrt{8+1} = \sqrt{9} = 3$.
If $f(x) = 5$, then $\sqrt{x+1} = 5 \Rightarrow x + 1 = 25 \Rightarrow x = 24$.

The domain is all the numbers that can go in, which is all real numbers ≥ -1 (you can't take the square root of a negative number unless we're dealing with imaginary numbers – not in the syllabus!). You write this as $\{x \geq -1, x \in \mathbb{R}\}$.

The range is all the numbers that can come out, which is all real numbers ≥ 0 (when you take a square root, the answer is always positive). You write this as $\{y \geq 0, y \in \mathbb{R}\}$.

How can my calculator help me do this?

Here your calculator is very, very handy! You can do all of the above using your GDC:

Evaluate a function at a point (example: $f(x) = x^2 - 3x$ at $x = 1$):

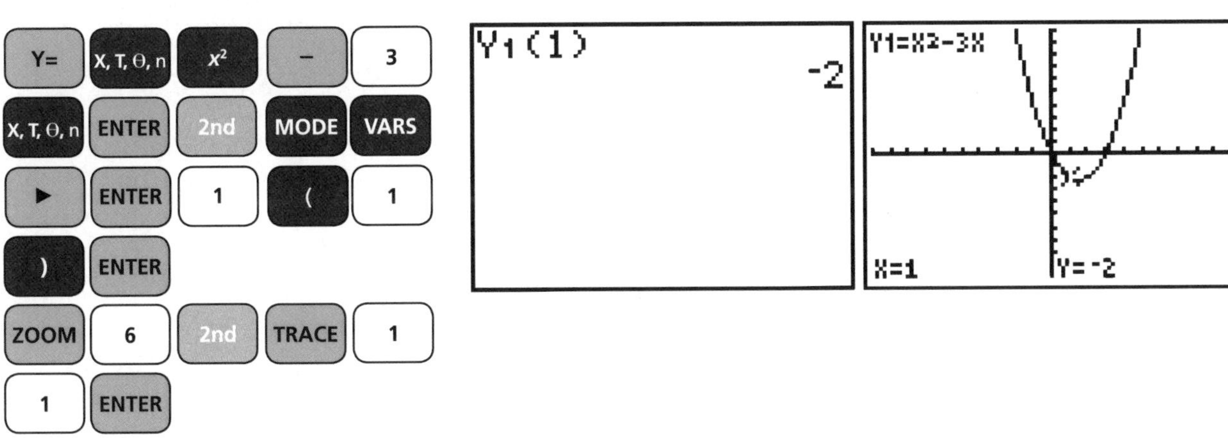

Find x given $f(x)$ (example: $f(x) = x^2 - 3x$ at $f(x) = -2$):

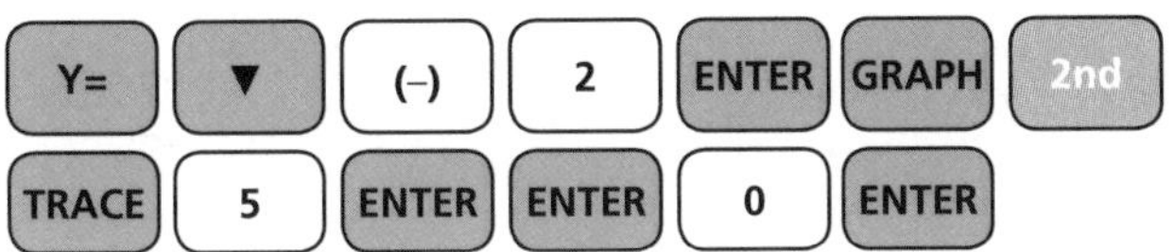

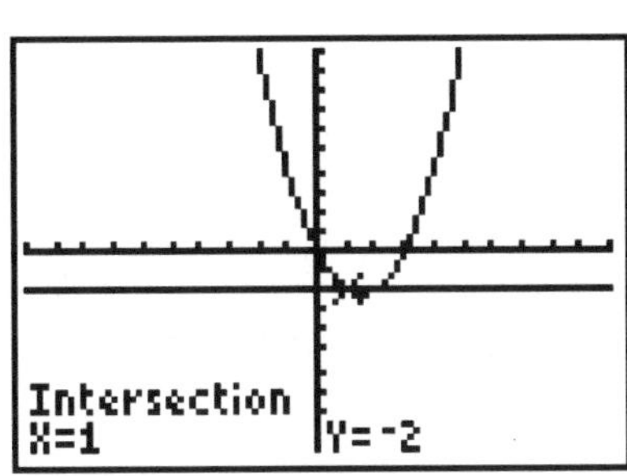

Find the domain and range (example: $f(x) = \sqrt{x+1}$):

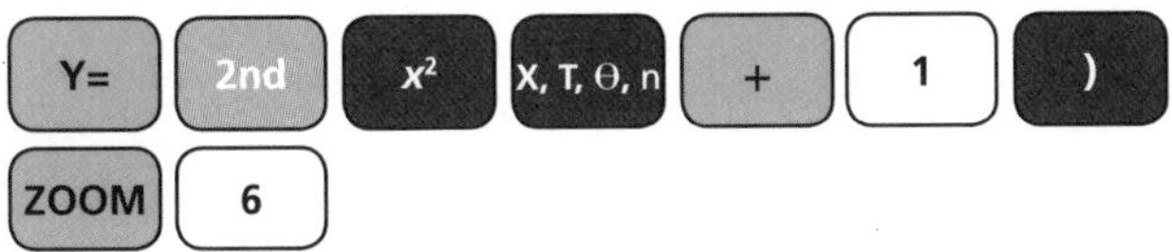

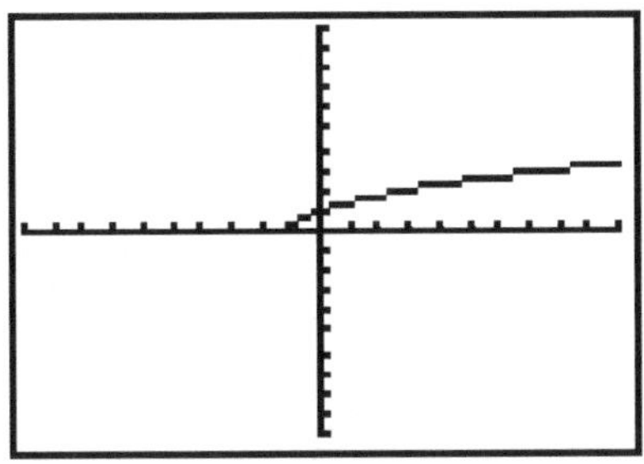

Look at the graph: the graph starts on the x-axis at $x = -1$ so the domain is $x \geq -1$; the graph starts on the y-axis at $y = 0$ so the range is $y \geq 0$.

Examples from past IB papers

An easy one from May 1996 Paper 1 (modified)

The cost C (in dollars) of posting a parcel of mass m (in grams) can be modelled by the function $C(m) = 0.064m + 0.040$.

(a) Calculate the cost of posting a parcel which has a mass of 40 grams.

(b) Calculate the mass of a parcel that costs $3.20 to post.

(c) State the domain and the range of this function.

Working:

(a) $C = 0.064(40) + 0.040$

$= 2.6$

(b) $3.20 = 0.064m + 0.040$

$3.16 = 0.064m$

$49.375 = m$

Answers:

(a) $2.60

(b) 49.4g (3sf)

(c) Domain: $m > 0$

Range: $c > 0.040$

A harder one from May 2007 Paper 1

The mapping below is of the form $f : x \mapsto a \times 2^x + b$ and maps the elements of x to elements of y.

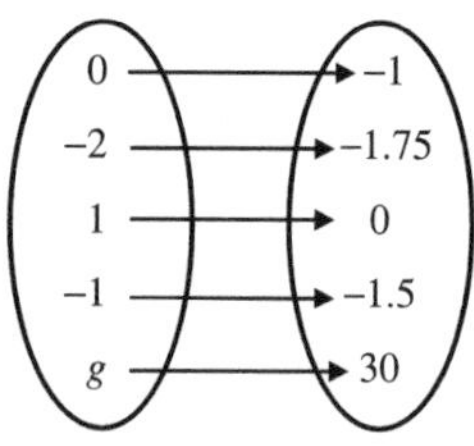

(a) (i) List the elements in the domain of f.
(ii) List the elements in the range of f.

(b) Find a and b.

(c) Find the value of g.

Working:

(b) $f(0) = -1$ so $a \times 2^0 + b = -1$ -> $a + b = -1$

$f(1) = 0$ so $a \times 2^1 + b = 0$ -> $2a + b = 0$

$-a = -1$

$a = 1$

$b = -2$

(c) $f(x) = (a)(2^x) + b$

$= 2^x - 2$

$f(g) = 2^g - 2 = 30$

$2^g = 32$

$g = 5$

Answers:

(a) (i) $-2, -1, 0, 1, g$

(ii) $-1.75, -1.5, -1, 0, 30$

(b) $a = 1, b = -2$

(c) 5

Now you practise it

An easy one from Specimen 2005 Paper 1

(a) Represent the function $y = 2x^2 - 5$ where $x \in \{-2, -1, 0, 1, 2, 3\}$ by a mapping diagram.

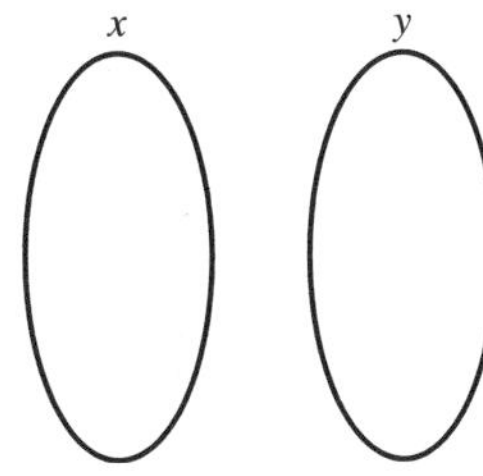

(b) List the elements of the domain of this function.

(c) List the elements of the range of this function.

Working:

Answers:

(b) ....

(c) ....

A harder one from November 2004 Paper 1

Write down the domain and range of the following functions.

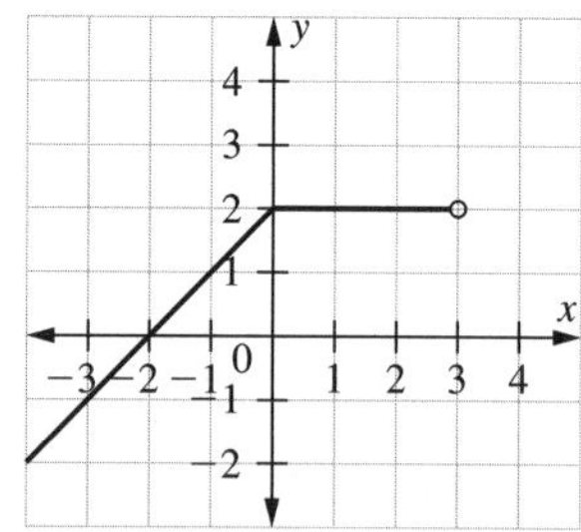

(b)

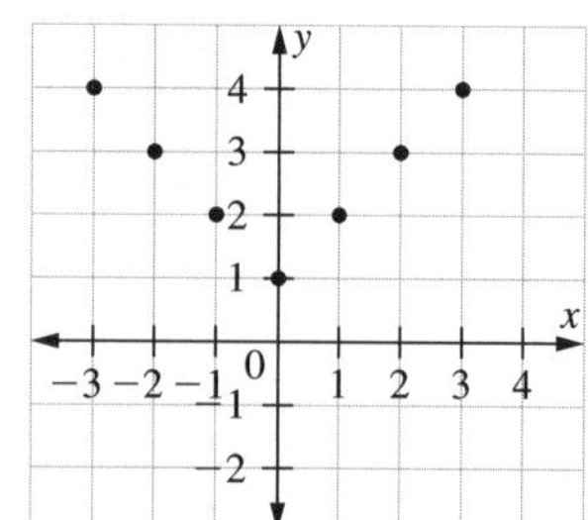

Working:

Answers:

(a) ....

....

....

(b) ....

TOPIC 14

Linear functions and graphs

What do you need to know?

The basic equation of a line is $y = mx + c$ where m is the slope/gradient of the line and c is the y-intercept. Another form is $ax + by + d = 0$ where $\frac{-d}{a}$ is the x-intercept and $\frac{-d}{b}$ is the y-intercept. You should know how to convert equations from one form to another and to clear out fractions in the second form so that a, b and d are integers.

SCOTT SAYS:

A good way to find the x- and y-intercepts of an equation in the form $ax + by + d = 0$ is to play "peek-a-boo". If you want to find the x-intercept, cover up the y term with your finger (hence "peek-a-boo") and solve the equation you see for x. Cover up the x term to find the y-intercept. It works!

How do you do it?

To graph a line in the form $y = mx + c$, you can either make a table of points (very slow) or simply plot a point at $(0, c)$ and then "count boxes" using the given slope/gradient. For example, if the slope/gradient is $\frac{3}{4}$,you would go up three, then go four to the right and plot another point.

To graph a line in the form $ax + by + d = 0$, calculate the x- and y-intercepts (either using the above formulae or by playing peek-a-boo), plot them on a graph and draw a line through the points. To clear out fractions, find the common denominator and multiply *everything* by this number.

To find an equation of a line, first find the slope/gradient (either given or using the slope formula $\frac{y_2 - y_1}{x_2 - x_1}$). Put the slope/gradient m into $y = mx + c$ and then substitute the coordinates of a point into x and y. Solve for c and write your equation. Convert into $ax + by + d = 0$ if necessary.

Examples

1. Given the equation of the line $y = 0.3x - 4$, the slope/gradient of the line is 0.3 or $\frac{3}{10}$. The y-intercept is -4.

   To graph, plot a point at $(0, -4)$, go up 3 and over 10 and then plot the next point; it should be at $(10, -1)$.

   To convert to $ax + by + d = 0$, move the x and the -4 to the other side: $-0.3x + y + 4 = 0$. To make these integers, multiply through by 10: $-3x + 10y + 40 = 0$.

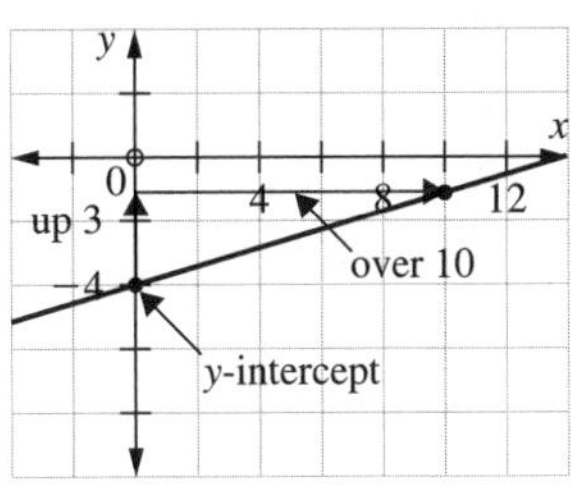

2. Given the equation of the line $x - 4y + 12 = 0$, the x-intercept is $\frac{-12}{1} = -12$ and the y-intercept is $\frac{-12}{-4} = 3$. To graph, plot the points $(-12, 0)$ and $(0, 3)$ and draw the line.

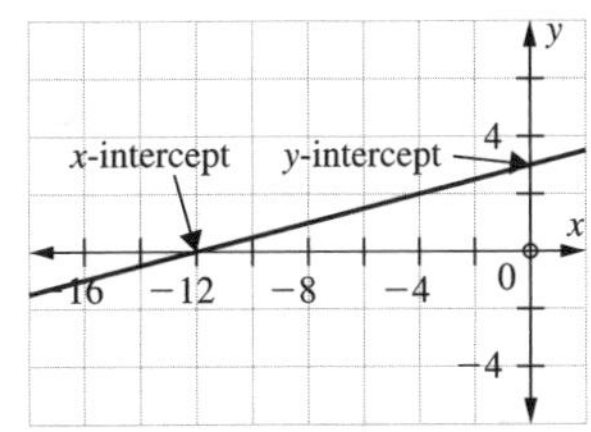

3. Given a line with gradient $\frac{-2}{7}$ and which goes through the point (3, 4), the equation of the line can be found. Substituting $\frac{-2}{7}$ for m into $y = mx + c$ gives $y = \frac{-2}{7}x + c$.
   Now substitute (3, 4) into the equation:

   $4 = \frac{-2}{7} \cdot 3 + c$

   $4 = \frac{-6}{7} + c$

   $c = 4 + \frac{6}{7} = \frac{28}{7} + \frac{6}{7} = \frac{34}{7}$

   so the equation is $y = \frac{-2}{7}x + \frac{34}{7}$.

How can my calculator help me do this?

You can, of course, graph any linear equation using your GDC if it is in the form $y = mx + c$. Just go to Y =, type your equation and then ZOOM 6 when you are done. If you want to graph an equation in the form $ax + by + d = 0$, you first have to convert it to $y = mx + c$.

If you want to find the x-intercept of a line (example: $y = 0.3x - 4$):

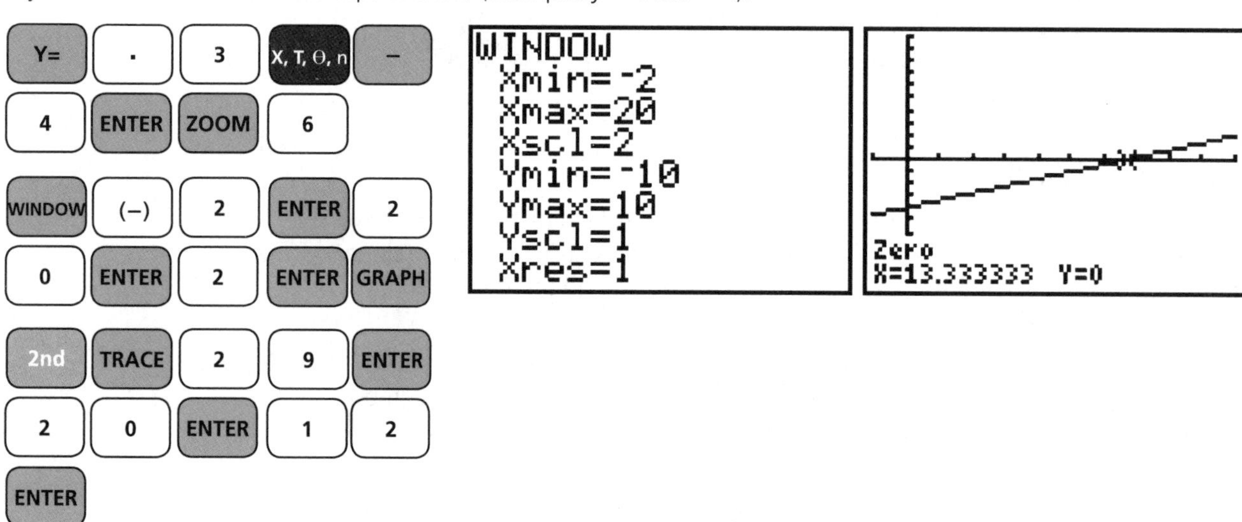

Notice that the window needed to be changed in order to see the intercept clearly!

Examples from past IB papers

An easy one from May 1997 Paper 1

An object moving in a straight line passes through the points (3, 0) and (2, 2). The object also passes through the point (0, a).

Calculate the value of a.

Working:

$y = mx + c$

(3, 0): $0 = 3m + c$

(2, 2): $2 = 2m + c$

$-2 = m$

$6 = c$

$y = -2x + 6$

(0, a): $a = -2(0) + 6$

$a = 6$

Answer:

(a) 6

A harder one from May 1998 Paper 1 (modified)

Two functions f and g are defined as follows:

$$f: x \mapsto 3x + 5$$
$$g: x \mapsto \frac{1}{2}(x - 4)$$

Find

(a) the x-intercept of f;

(b) the y-intercept of g;

(c) the coordinates of the point of intersection of f and g.

Working:

(a) $3x + 5 = 0$

$x = \frac{-5}{3}$

(b) $y = \frac{1}{2}(0 - 4)$

$y = -2$

(c) $3x + 5 = \frac{1}{2}(x - 4)$

$3x + 5 = \frac{1}{2}x - 2$

$\frac{5}{2}x = -7$

$x = \frac{-14}{5}$

$y = 3\left(-\frac{14}{5}\right) + 5$

$= -\frac{17}{5}$

Answers:

(a) $x = \frac{-5}{3}$

(b) $y = -2$

(c) $\left(-\frac{14}{5}, -\frac{17}{5}\right)$

Now you practise it

An easy one from May 2005 Paper 1

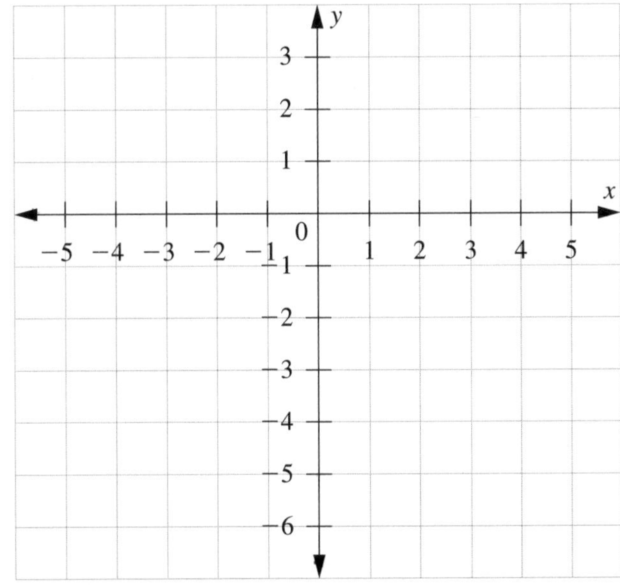

(a) On the grid above, draw a straight line with a gradient of −3 that passes through the point (−2, 0).

(b) Find the equation of this line.

Working:

Answer:

(b) ....

A harder one from May 2007 Paper 1

A shop sells bread and milk. On Tuesday, 8 loaves of bread and 5 litres of milk were sold for \$21.40. On Thursday, 6 loaves of bread and 9 litres of milk were sold for \$23.40.

If b = the price of a loaf of bread and m = the price of one litre of milk, Tuesday's sales can be written as $8b + 5m = 21.40$.

(a) Using simplest terms, write an equation in b and m for Thursday's sales.

(b) Find b and m.

(c) Draw a sketch, in the space provided, to show how the prices can be found graphically.

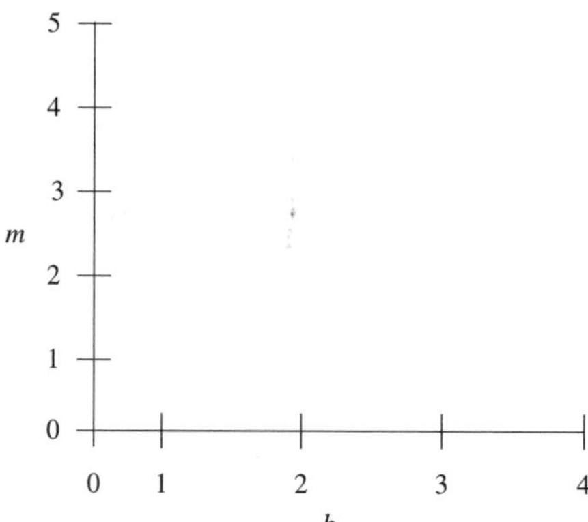

Working:

Answers:

(a) ....

(b) ....

IRENE SAYS:

Aha! Did you notice that this is a combined question? The first part is on systems of linear equations and then part (c) is on graphing linear functions. Tricky!

TOPIC 15

Quadratic functions and graphs

What do you need to know?

The basic form of a quadratic function is $y = ax^2 + bx + c$ where a, b and c are numbers. The graph of a quadratic equation is always a *parabola* – a bowl-shaped curve that can either open upwards (if a is positive) or downwards (if a is negative).

The vertex of a parabola is the bottom/top of the curve and you find it using the formula $-\frac{b}{2a}$. This is in your formula booklet under Topic 4 – Functions:

4.3	Equation of axis of symmetry	$x = -\frac{b}{2a}$

The axis of symmetry is the vertical line that goes through the vertex. Its equation is always in the form $x = n$ and n is the x-coordinate of the vertex.

How do you do it?

To graph a parabola from an equation, you can make a table of points (use your GDC to do this quickly) and/or learn some useful tricks:

- Factorise the equation and set each factor equal to zero. This will give you the two x-intercepts quickly. Find the middle and you can draw an axis of symmetry.
- If the equation is in the form $ax^2 + bx + c$, then the x-coordinate of the vertex is $-\frac{b}{2a}$. Put that number into the equation to get the y-coordinate. Draw an axis of symmetry and "mirror" points from one side to the other.
- If the equation is in the form $ax^2 + bx + c$, c is the y-intercept.

Example

To graph the quadratic equation $y = x^2 - 6x + 2$, you can make a table of points:

x	−2	−1	0	1	2	3	4	5	6	7
y	18	9	2	−3	−6	−7	−6	−3	2	9

Notice that the vertex must be at (3, −7), and the equation of the axis of symmetry is thus $x = 3$.

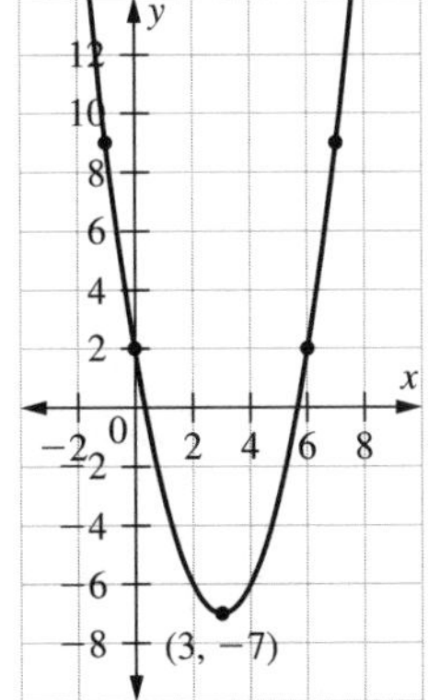

To find the vertex easily, $-\frac{b}{2a}$ becomes $-\frac{(-6)}{2.1} = \frac{6}{2} = 3$ = the x-coordinate of the vertex. The y-coordinate then is $3^2 - 6 \cdot 3 + 2 = -7$. So the vertex is (3, −7).

The y-intercept is just (0, 2) because $c = 2$.

How can my calculator help me do this?

How to get a table of points of the quadratic equation $y = x^2 - 6x + 2$:

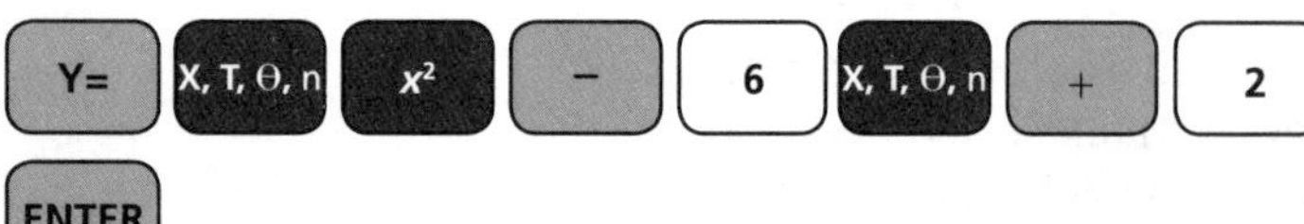

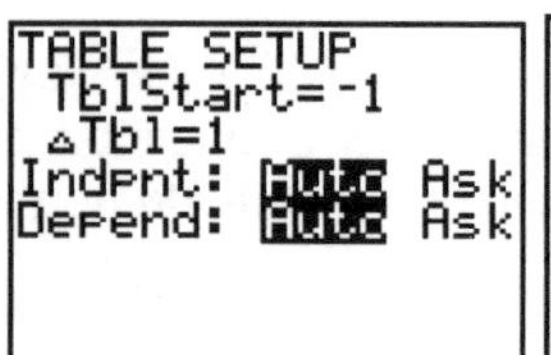

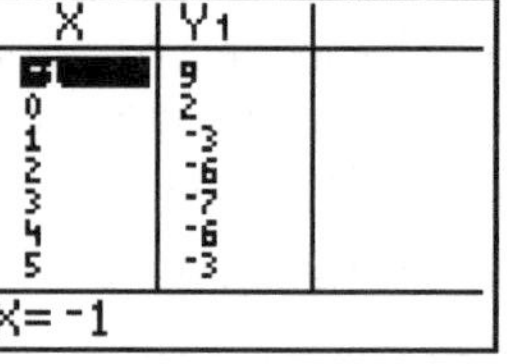

ENTER

2nd GRAPH

How to find the vertex of the quadratic equation $y = x^2 - 6x + 2$:

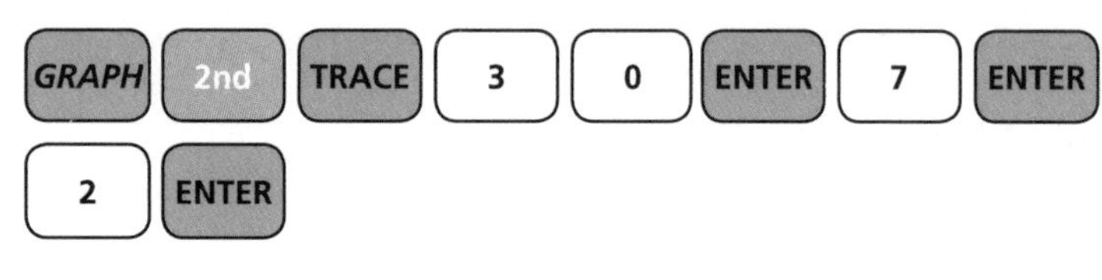

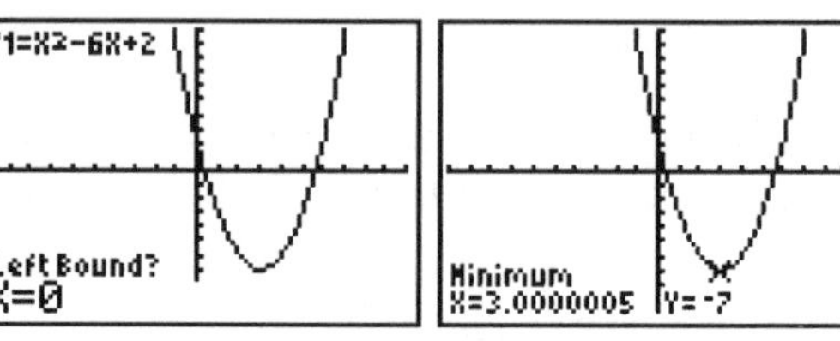

Examples from past IB papers

An easy one from November 2003 Paper 1

The graph of the function $f : x \mapsto 30x - 5x^2$ is given in the diagram below.

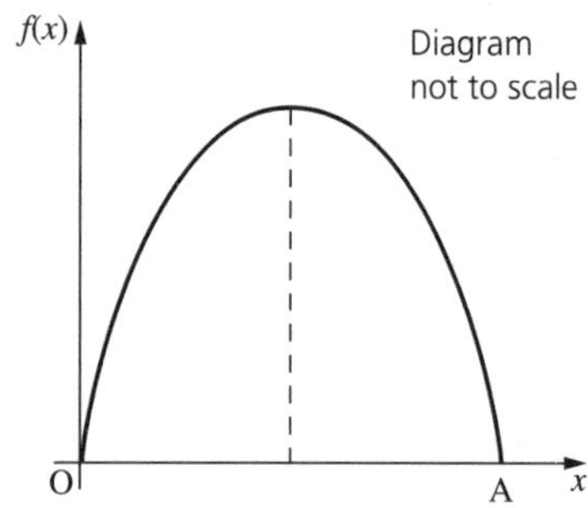

(a) Factorise fully $30x - 5x^2$.

(b) Find the coordinates of the point A.

(c) Write down the equation of the axis of symmetry.

Working:

(a) $30x - 5x^2$

$= 5(6x - x^2)$

$= 5x(6 - x)$

(b) $5x = 0$, $6 - x = 0$

$x = 0$, $6 = x$ -> point A

(c) $x = \frac{-b}{2a} = \frac{-30}{(2X-5)} = 3$

$x = 3$

Answers:

(a) $5x(6 - x)$

(b) $(6, 0)$

(c) $x = 3$

A harder one from November 2006 Paper 1

The graph of a quadratic function $f(x)$ intersects the horizontal axis at (1, 0) and the equation of the axis of symmetry is $x = -1$.

(a) Write down the x-coordinate of the other point where the graph of $y = f(x)$ intersects the horizontal axis.

(b) $y = f(x)$ reaches its maximum value at $y = 5$.

(i) Write down the value of $f(-1)$.

(ii) Find the range of the function $y = f(x)$.

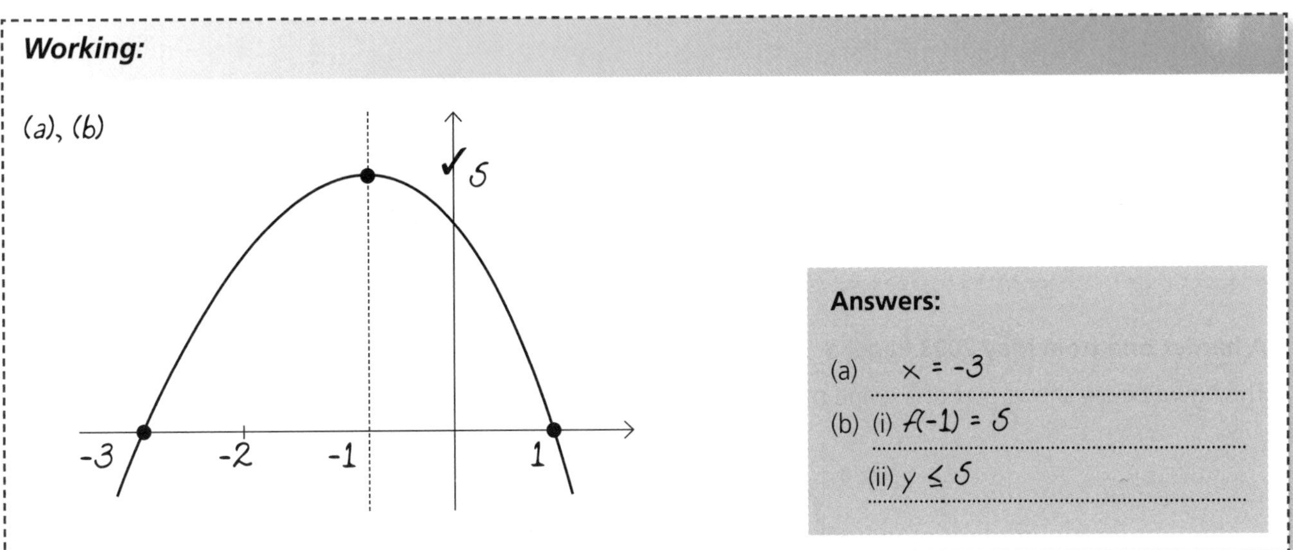

Now you practise it

An easy one from November 2004 Paper 1

The graph of the function $f(x) = x^2 - 2x - 3$ is shown in the diagram below.

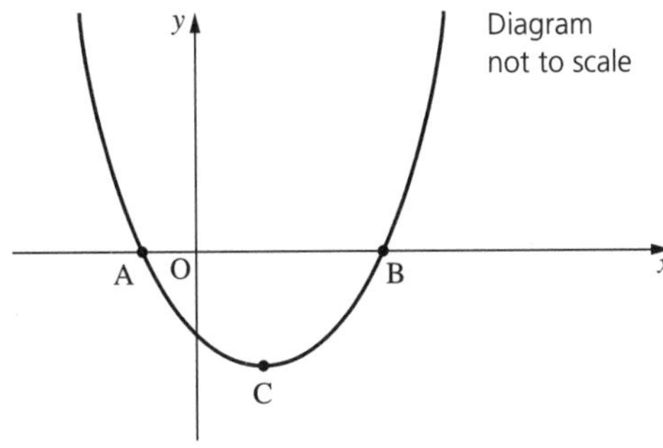

(a) Factorise the expression $x^2 - 2x - 3$.

(b) Write down the coordinates of the points A and B.

(c) Write down the equation of the axis of symmetry.

(d) Write down the coordinates of the point C, the vertex of the parabola.

Working:

Answers:

(a) ....

(b) ....

(c) ....

(d) ....

A harder one from May 2003 Paper 1

The figure below shows part of a graph of a quadratic function $y = ax^2 + 4x + c$.

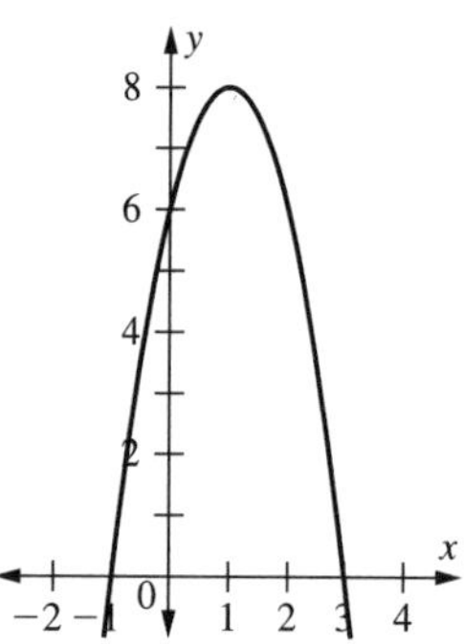

(a) Write down the value of c.

(b) Find the value of a.

(c) Write the quadratic function in its factorised form.

Working:

Answers:

(a) ....

(b) ....

(c) ....

TOPIC 16

Exponential functions and graphs

What do you need to know?

The basic form of an exponential function is $f(x) = a \cdot b^x$, where a is the *initial amount* and b is the *exponential rate*. All exponential functions in this form go through the point (0,1) and have an asymptote on the x-axis (if there is a number at the end like +2 or −3 then the graph goes up or down by that amount).

If b is larger than 1, then it is exponential growth and looks like this:

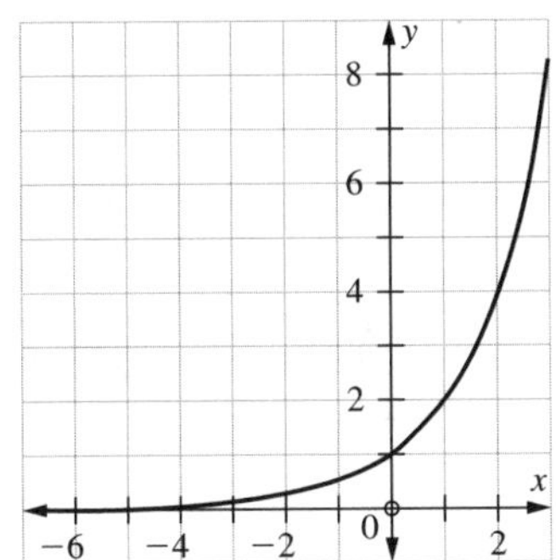

If b is between 0 and 1, then it is exponential decay and looks like this:

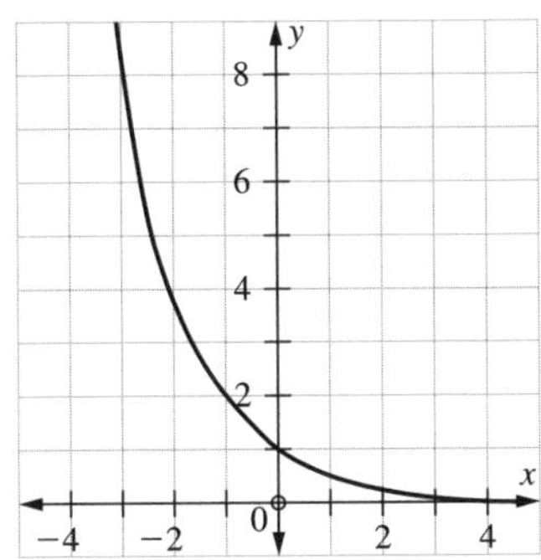

How do you do it?

IB questions on exponential functions usually ask you to graph the function, identify parts of the graph (initial values, asymptotes), substitute values in the function or find values of x.

To graph the function, make a table of points (like with other functions) and draw a smooth curve through them. Don't forget to label the y-intercept and the asymptote! The equation of the asymptote will be in the form $y =$.

To substitute values, you just put the value of x into the function. Sometimes you are given $f(x)$ and you have to find x; this is harder – you have to do it graphically! (You can also find x using algebra but it involves logarithms. If you have studied logarithms, go ahead and use them but they are not on the syllabus.) Look at the example below.

Example

A particular cancerous tumour grows exponentially from diagnosis in a way that is modelled by the function $M(t) = 1.2 \cdot 1.06^t$, where M is the mass in grams and t is the time in days. The initial mass is 1.2 grams and the growth rate is 1.06 (6% growth). A table of points can be made:

t	0	5	10	15	20	25	30
M	1.2	1.61	2.15	2.88	3.85	5.15	6.89

And a graph can now be created:

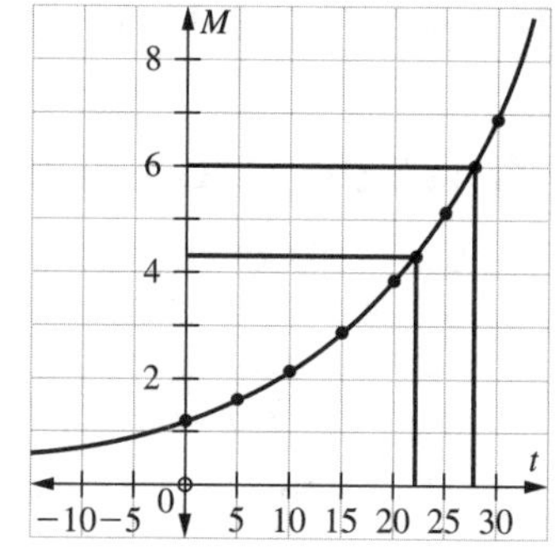

If you want to find the mass after 22 days, go to 22 on the t-axis and read off where it hits the graph: around 4.3 grams.

If you want to find how long it will take to reach 6 grams, go to 6 on the M-axis and see where it hits the graph: around 27.6 days.

How can my calculator help me do this?

Exponential function problems are perfect for the GDC! Let's go through the whole previous example. Note that the hardest part about doing exponential functions on the GDC is the *window*! Plan ahead and leave yourself lots of room…

How to get a table of points of the exponential equation $y = 1.2 \cdot 1.06^x$:

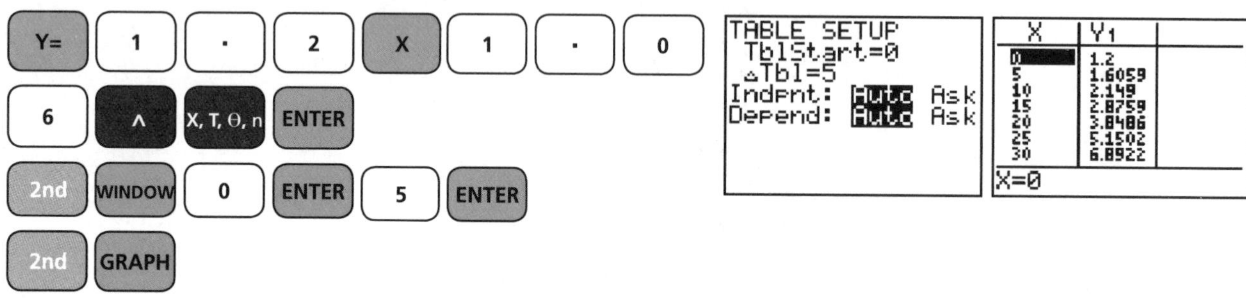

How to make a graph and find the mass after 22 days:

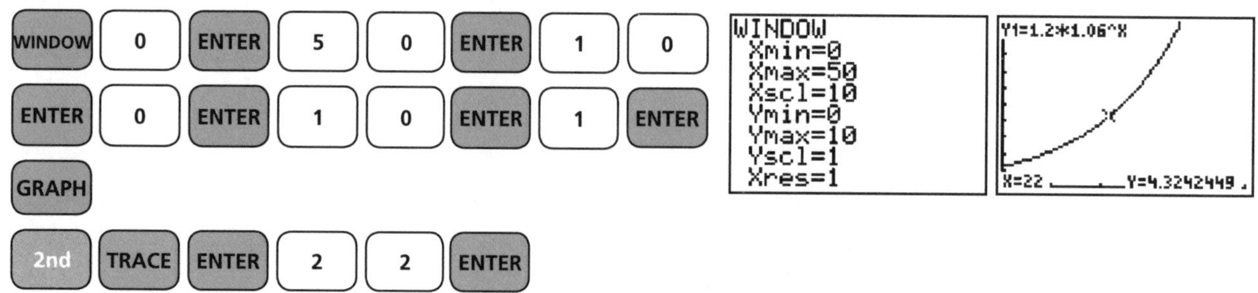

How to find how long it takes to reach 6 grams:

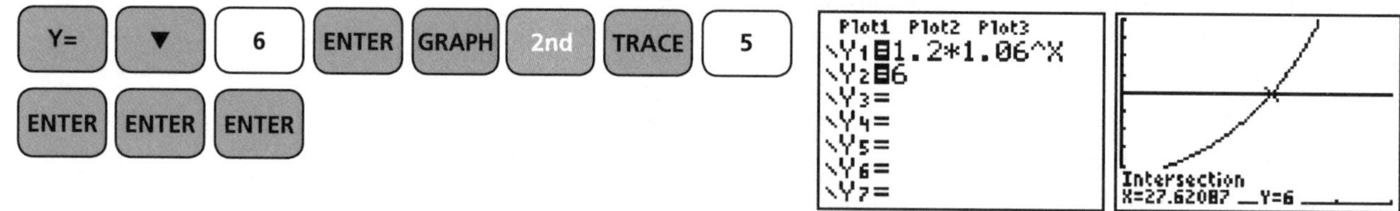

Examples from past IB papers

An easy one from November 2000 Paper 1

The following diagram shows part of the graph of an exponential function $f(x) = a^{-x}$, where $x \in \mathbb{R}$.

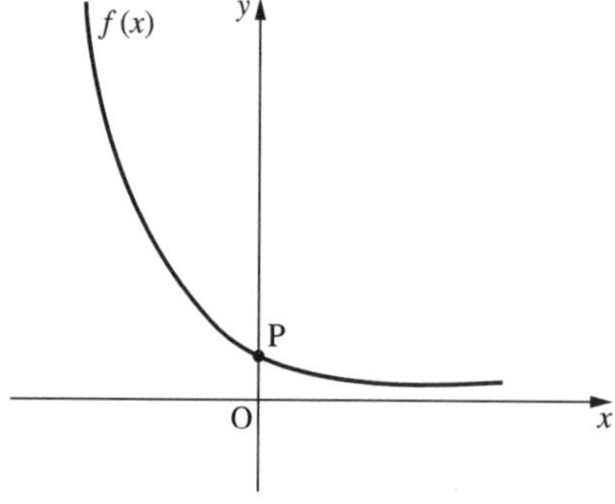

(a) What is the range of f?

(b) Write down the coordinates of the point P.

(c) What happens to the values of $f(x)$ as elements in its domain increase in value?

Working:

(b) $f(x) = a^{-x}$

$f(0) = a^0$

$f(0) = 1$

Answers:

(a) positive real numbers (R^+)

(b) (0, 1)

(c) decreases towards zero

A harder one from May 2003 Paper 1

The diagram below shows a chain hanging between two hooks A and B. The points A and B are at equal heights above the ground. P is the lowest point on the chain. The ground is represented by the x-axis. The x-coordinate of A is -2 and the x-coordinate of B is 2. Point P is on the y-axis. The shape of the chain is given by

$$y = 2^x + 2^{-x}, \text{ where } -2 \le x \le 2.$$

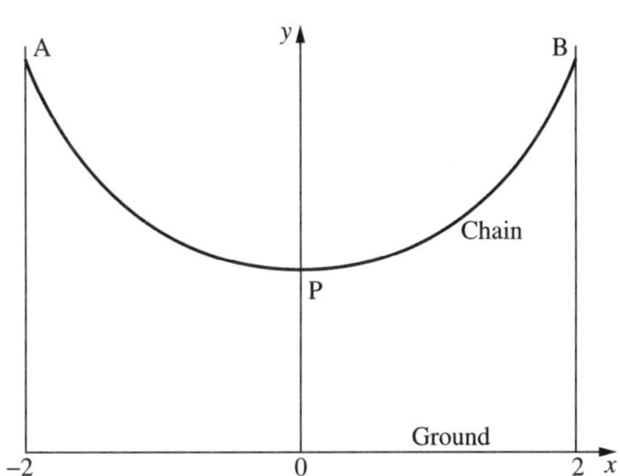

(a) Calculate the height of the point P.

(b) Find the range of y. Write your answer as an interval or using inequality symbols.

Working:

(a) Point P is at $x = 0$.

$y(0) = 2^0 + 2^{-0}$

$= 1 + 1 = 2$

(b) (0, 2) is lowest point.

$x = 2$ and $x = -2$ are highest points.

At $x = 2$, $y = 2^2 + 2^{-2} = 4\frac{1}{4}$

Highest points are

$(2, 4\frac{1}{4})$ and $(-2, 4\frac{1}{4})$

Answers:

(a) 2

(b) $2 \le y \le 4\frac{1}{4}$

Now you practise it

An easy one from November 2006 Paper 1

The value of a car decreases each year. This value can be calculated using the function $v = 32\,000r^t$, $t \geq 0$, $0 < r < 1$, where v is the value of the car in USD, t is the number of years after it was first bought and r is a constant.

(a) (i) Write down the value of the car when it was first bought.

(ii) One year later the value of the car was 27 200 USD. Find the value of r.

(b) Find how many years it will take for the value of the car to be less than 8000 USD.

Working:

Answers:

(a) (i) ....

(ii) ....

(b) ....

A harder one from November 2001 Paper 1

The graph below shows the curve $y = k(2^x) + c$, where k and c are constants.

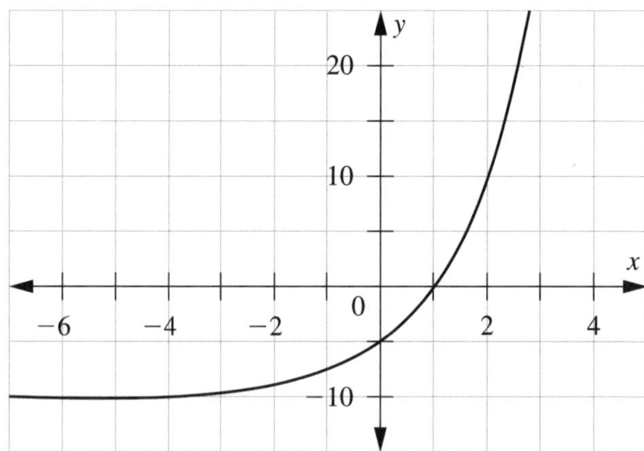

Find the values of c and k.

Working:

Answer:

....

Trigonometric functions and graphs

What do you need to know?

Trigonometric functions for Math Studies SL only include graphs of sin and cos. The only forms you need to know are:

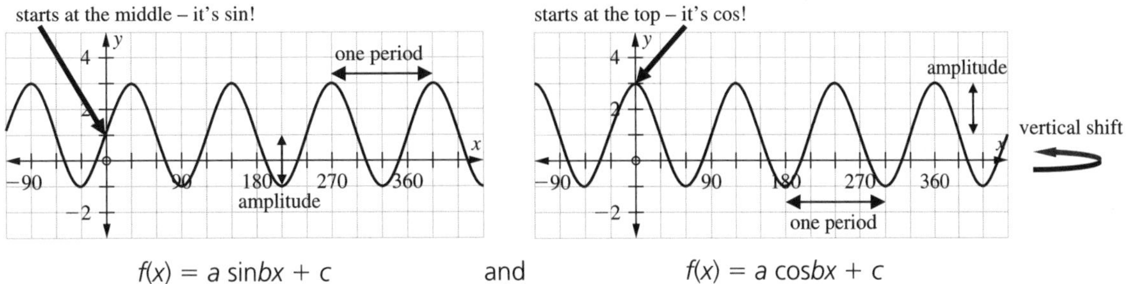

$f(x) = a \sin bx + c$ and $f(x) = a \cos bx + c$

where a, b and c can be integers, fractions, decimals, etc… The x-axis is in degrees and the y-axis is just normal numbers.

SCOTT SAYS:

To tell whether a graph is sine or cosine, *look at the y-intercept*. If the graph starts at the top or bottom of its cycle, it's cosine. If it starts in the middle, it's sine.

The *amplitude* of a trig function is the distance from the middle of the curve to either the top or the bottom.

The *period* of a trig function is the length of one complete cycle.

How do you do it?

You find the amplitude either by looking at the value of a (but remember that the amplitude of a trig function must always be positive – take the absolute value of a) or by looking at the graph.

You find the period either by dividing 360 by the b value (i.e. the period $= \frac{360}{b}$) or again by looking at the graph.

You find the c value (called the *vertical shift*) by looking at the middle of the graph and seeing how far it is from the x-axis.

HANA AND CHRIS SAY:

Be careful! When your IB Coordinator resets your calculator for the IB exam, it will be reset back to *radian measure* for trig. As soon as you are allowed to use your calculator again, make sure to go to MODE and put your calculator back into degrees!

Examples

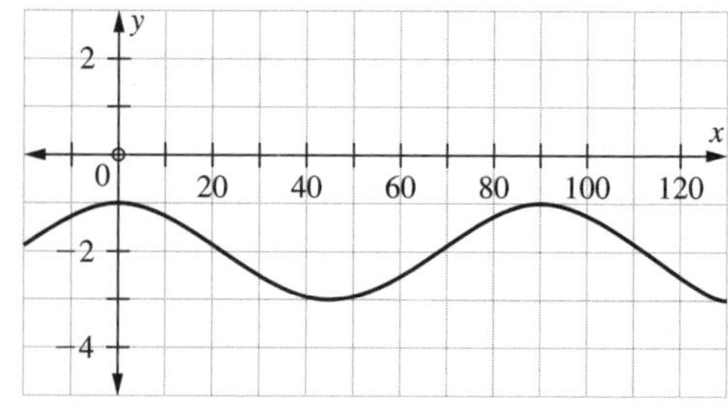

1. Given the graph on the right, you can see that
   it starts at the top (so it's cos),
   the distance from the top to the middle is 1 (so the amplitude $= a = 1$),
   one cycle is 90° (so the period is 90°, and thus $b = 4$),
   the middle of the graph is shifted down 2 from the x-axis (so $c = -2$).

   Hence the equation is $y = 1 \cos 4x - 2$ or simply $y = \cos 4x - 2$.

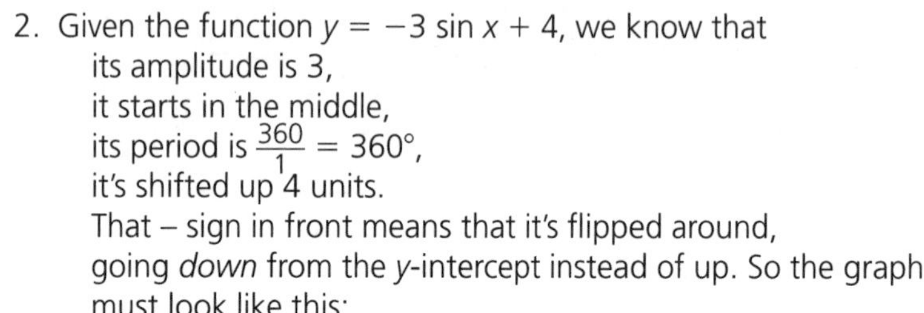

2. Given the function $y = -3 \sin x + 4$, we know that
   its amplitude is 3,
   it starts in the middle,
   its period is $\frac{360}{1} = 360°$,
   it's shifted up 4 units.
   That – sign in front means that it's flipped around, going *down* from the y-intercept instead of up. So the graph must look like this:

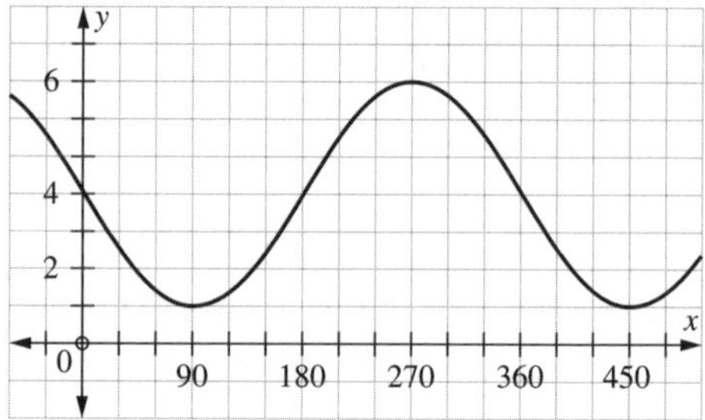

How can my calculator help me do this?

This is all similar to exponential functions except the window settings are different. And don't forget to make sure your calculator is in *degrees*!

How to graph the trig equation $y = \cos 4x - 2$:

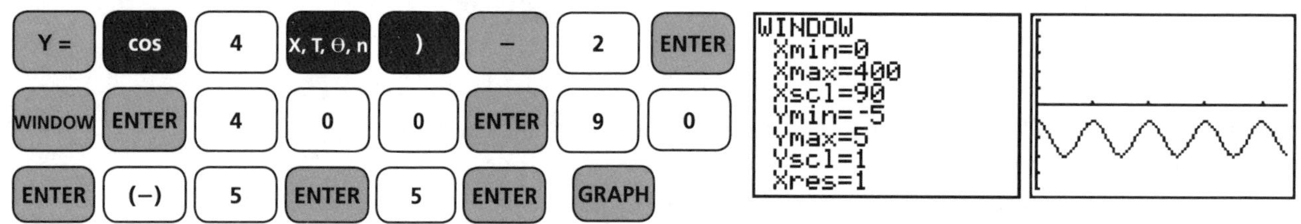

How to get a table of points:

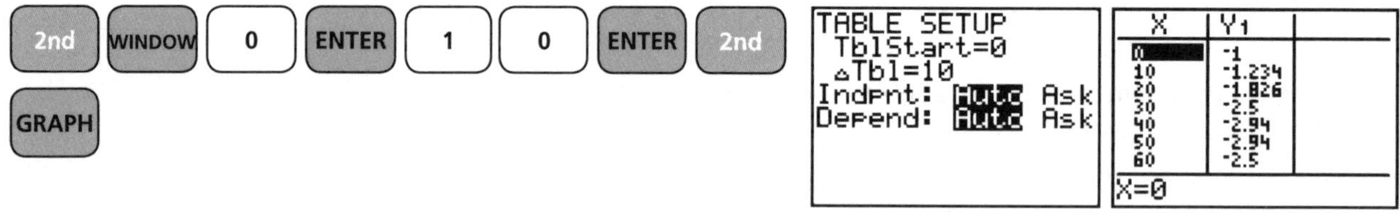

Examples from past IB papers

An easy one from May 1998 Paper 1

The temperature during a 24-hour period is illustrated on the graph and is given by the function $f : x \rightarrow a + 2 \sin bt$, where t is the time in hours.

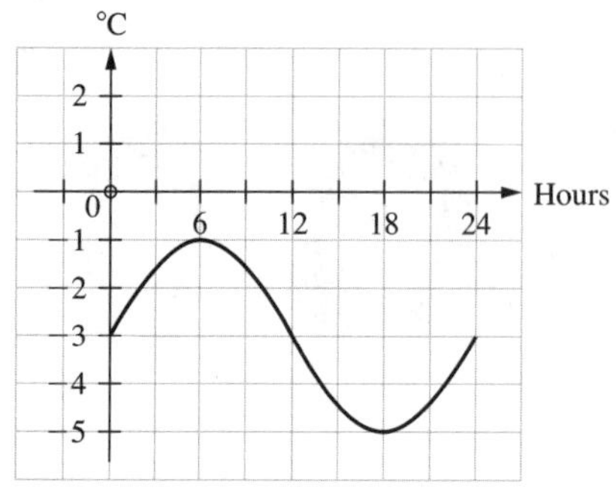

(a) From the graph, estimate the temperature after 15 hours.

(b) What is the value of *a*?

(c) What is the value of *b*?

Working:

(c) $\frac{360}{24} = b$

Answers:

(a) -4.5°C

(b) -3

(c) 15

A harder one from November 2006 Paper 1

The temperature (°C) during a 24-hour period in a certain city can be modelled by the function $T(t) = -3 \sin(bt) + 2$, where b is a constant, t is the time in hours and bt is measured in degrees. The graph of this function is illustrated below.

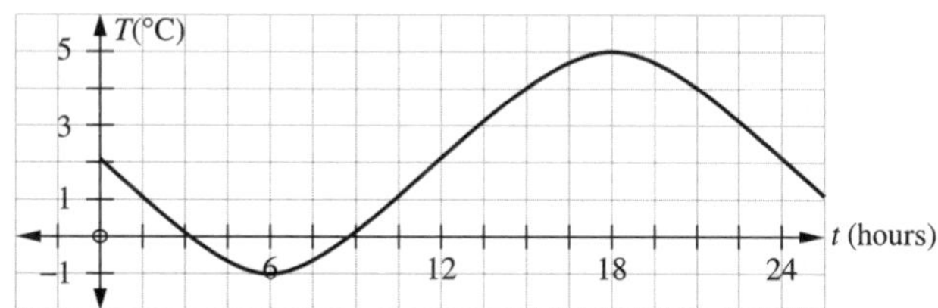

(a) Determine how many times the temperature is exactly 0°C during this 24-hour period.

(b) Write down the time at which the temperature reaches its maximum value.

(c) Write down the interval of time in which the temperature changes from −1°C to 2°C.

(d) Calculate the value of *b*.

Working:

(d) $\frac{360}{b} = 24$

$b = 15$

Answers:

(a) 2 times

(b) 18 hours

(c) $6 \leq t \leq 12$

(d) 15

Now you practise it

An easy one from November 2007 Paper 1

(a) Sketch the graph of the function $y = 1 + \frac{\sin(2x)}{2}$ for $0° \leq x \leq 360°$ on the axes below.

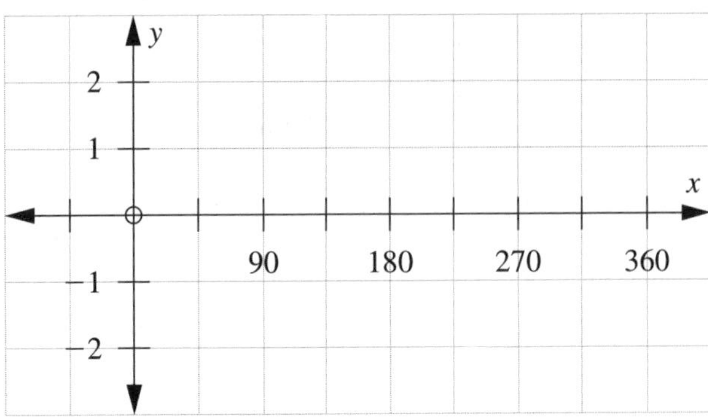

(b) Write down the period of the function.

(c) Write down the amplitude of the function.

Working:

Answers:

(b) ....

(c) ....

IRENE SAYS:

Did I trick you? I wrote the function in a weird way with the 1 + in front and the 2 down below. If you rewrote the function as $y = \frac{1}{2}\sin(2x) + 1$, well done!

A harder one from May 2007 Paper 1

The graph of $y = a\sin 2x + c$ is shown below, $-180 \leq x \leq 360$, x is measured in degrees.

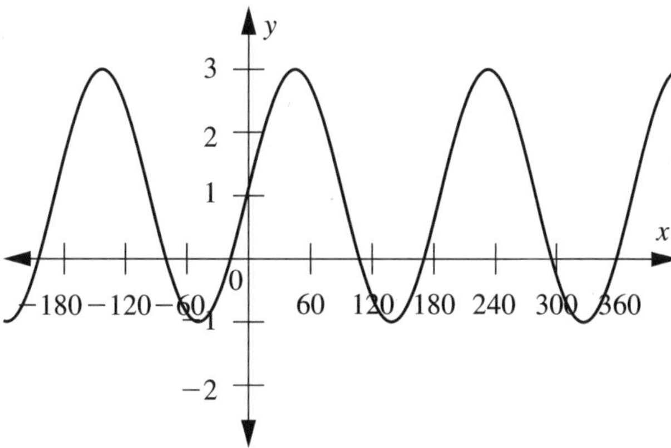

(a) State:

(i) the period of the function,

(ii) the amplitude of the function.

(b) Determine the values of a and c.

(c) Calculate the value of the first **negative** x-intercept.

Working:

Answers:

(a) (i) ....

(ii) ....

(b) ....

(c) ....

TOPIC 18

Unfamiliar functions and graphs

What do you need to know?

Unfamiliar functions in Math Studies SL are functions that you are not expected to have seen before in the course. What are you expected to know, then?? All you need to know is

- How to type a weird function into your calculator correctly, including the correct use of parentheses / brackets
- How to find on a calculator where two functions intersect (they will often set two functions equal to each other and ask you to find x)
- How to identify and find the equations for asymptotes in a graph (asymptotes are lines, usually vertical or horizontal in Math Studies, that a curve approaches but does not touch).

How do you do it?

This entire topic is calculator-based. You just type each function into your GDC, graph them, make sure the window is right, find the point(s) of intersection and answer the question. Easy!

Examples

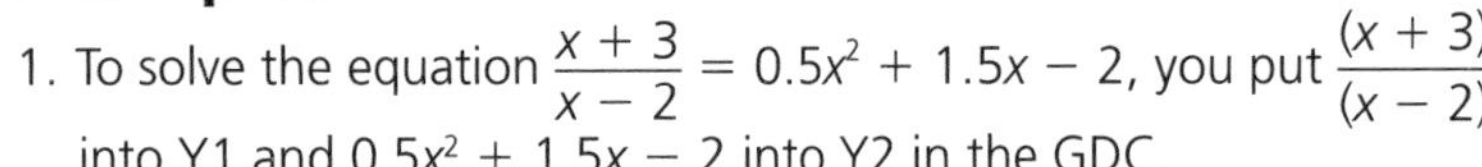

1. To solve the equation $\frac{x+3}{x-2} = 0.5x^2 + 1.5x - 2$, you put $\frac{(x+3)}{(x-2)}$ into Y1 and $0.5x^2 + 1.5x - 2$ into Y2 in the GDC.

   Then find the points of intersection: $(-4.07, 0.1762)$, $(0.1695, -1.731)$, $(2.9, 6.555)$.

   The y-coordinates are not relevant; we are solving for x. So the answers are $x = -4.07$ (3sf), $x = 0.170$ (3sf) and $x = 2.9$.

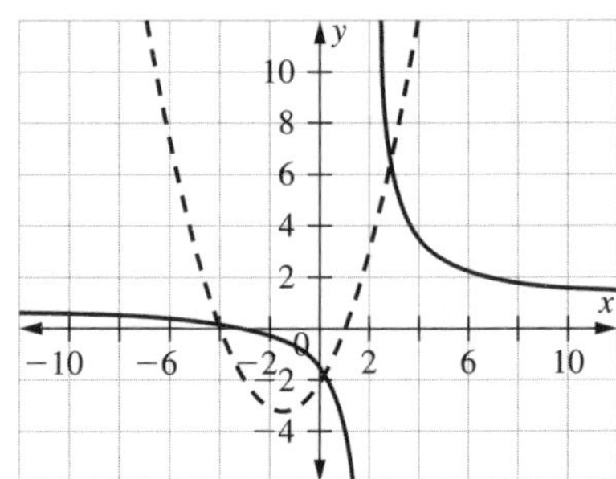

IRENE SAYS:

Make sure you round all your answers to 3sf! An "accuracy penalty" is given to answers that are not rounded correctly.

2. To find the equations of the asymptotes of the function $f(x) = \frac{x+3}{x-2}$, you have to look at the graph. Clearly there is both a horizontal and a vertical asymptote.

   To find the equation of the horizontal asymptote, it's easiest just to trace a bit to the right or left and watch the y-coordinates. They will get closer and closer to 1, so because it's a horizontal line, the equation is $y = 1$.

   For the vertical asymptote, you can either trace like before or just look at the denominator of the function. You know you can't divide by zero, so $x - 2 \neq 0$ or $x = 2$.

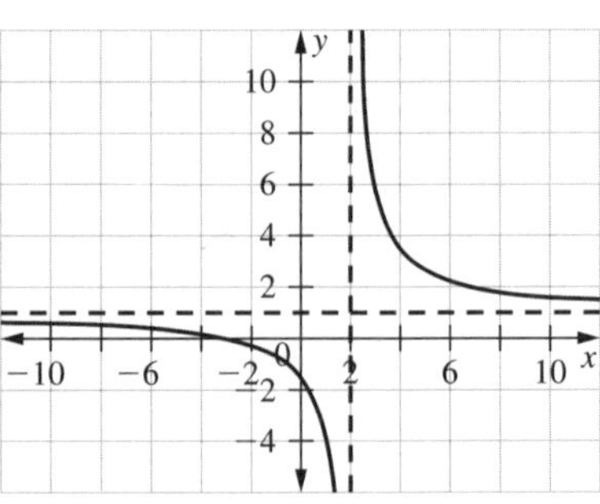

How can my calculator help me do this?

Again, this topic is **all** using the GDC. Lots of things to do here.

How to solve the equation $\frac{x+3}{x-2} = 0.5x^2 + 1.5x - 2$:

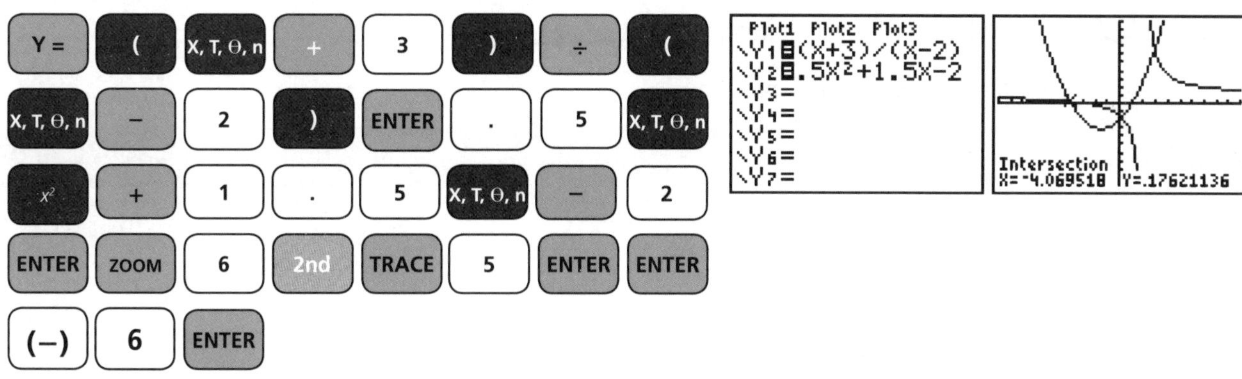

(continue to find the other two points of intersection)

How to find the asymptotes of the function $f(x) = \dfrac{x+3}{x-2}$ using a graph:

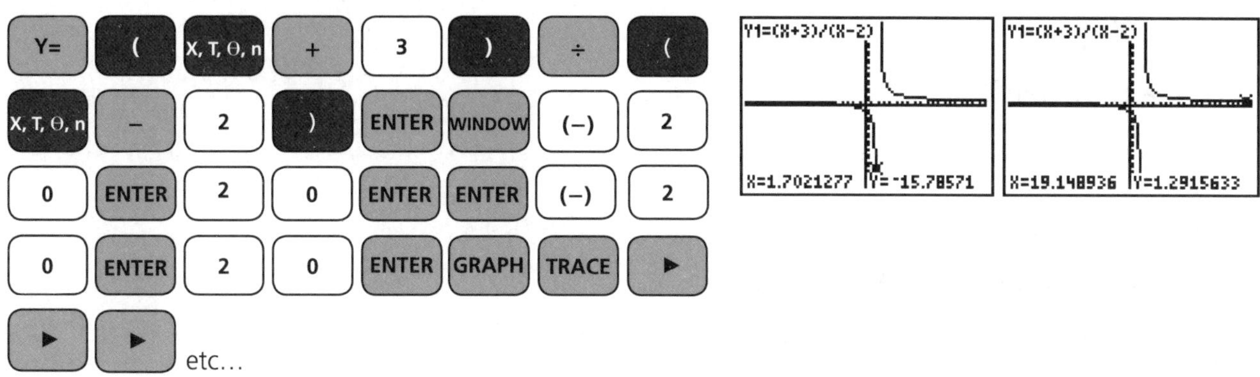

etc...

And continuing to find the horizontal asymptote of $f(x) = \dfrac{x+3}{x-2}$ using a table:

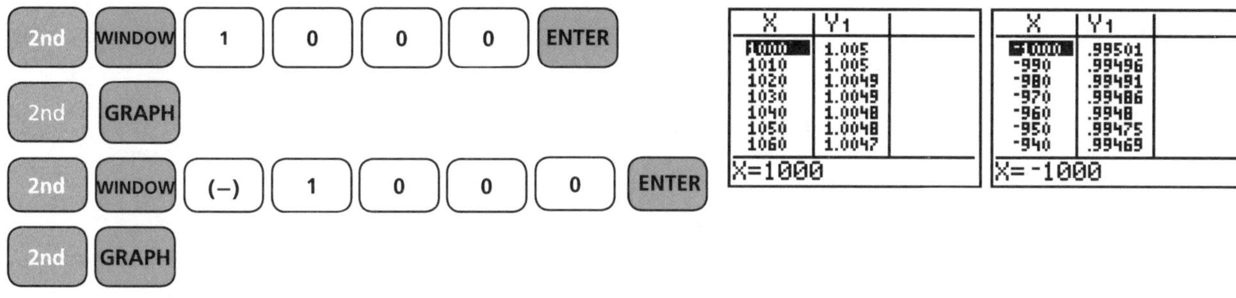

SCOTT SAYS:

Notice here how the y-coordinates get closer and closer to 1 in both cases – hence the equation of the asymptote is y = 1!

Examples from past IB papers

An easy one from May 2006 Paper 1

Two functions $f(x)$ and $g(x)$ are given by

$$f(x) = \frac{1}{x^2+1},$$
$$g(x) = \sqrt{x},\ x \geq 0.$$

(a) Sketch the graphs of $f(x)$ and $g(x)$ together on the same diagram using values of x between -3 and 3, and values of y between 0 and 2. You must label each curve.

(b) State how many solutions exist for the equation $\frac{1}{x^2+1} - \sqrt{x} = 0$.

(c) Find a solution of the equation given in part (b).

Working:

(a)

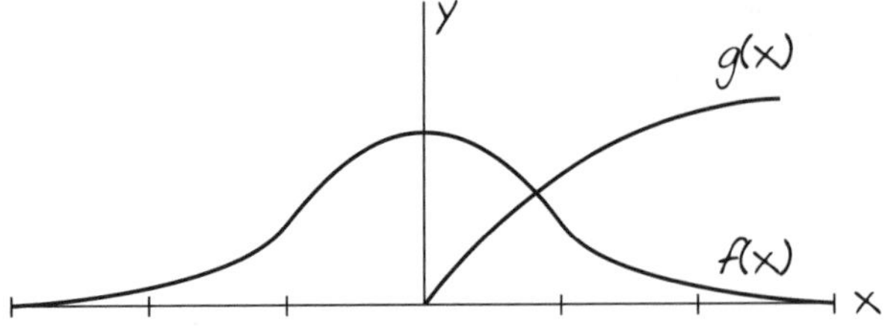

(b) *one point of intersection*

(c) *use GDC*

$x = 0.56984029$

Answers:

(b) *one*

(c) $x = 0.570$

Now you practise it

An easy one from Specimen 2005 Paper 1

(a) Sketch a graph of $y = \frac{x}{2+x}$ for $-10 \leq x \leq 10$.

(b) Hence write down the equations of the horizontal and vertical asymptotes.

Working:

Answer:

(b) ....

A harder one from November 2006 Paper 1

The figure below shows the graphs of the functions $f(x) = 2^x + 0.5$ and $g(x) = 4 - x^2$ for values of x between -3 and 3.

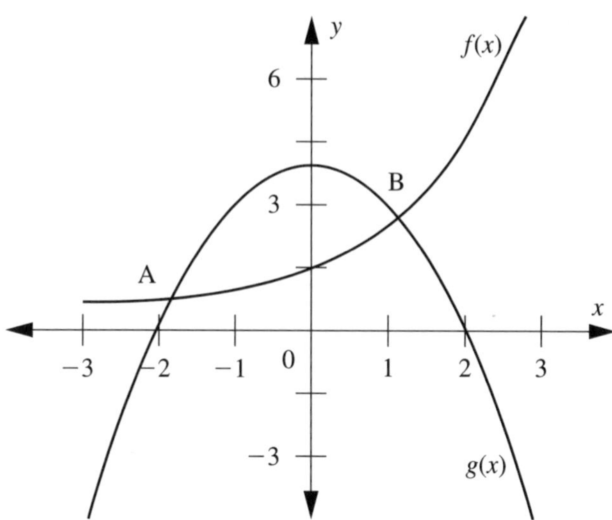

(a) Write down the coordinates of the points A and B.

(b) Write down the set of values of x for which $f(x) < g(x)$.

Working:

Answers:

(a) ....

(b) ....

Additional practice problems

1. From May 2003 Paper 2 *[Maximum mark: 14]*

The number of bacteria (y) present at any time is given by the formula:

$y = 15\,000e^{-0.25t}$, where t is the time in seconds and e = 2.72 correct to 3 s.f.

(a) Calculate the values of a, b and c to the nearest hundred in the table below: *[3 marks]*

Time in seconds (t)	0	1	2	3	4	5	6	7	8
Amount of bacteria (y) (nearest hundred)	a	11 700	9100	7100	b	4300	3300	2600	c

(b) On graph paper, using 1 cm for each second on the horizontal axis and 1 cm for each thousand on the vertical axis, draw and label the graph representing this information. *[5 marks]*

(c) Using your graph, answer the following questions:

(i) After how many seconds will there be 5000 bacteria? Give your answer correct to the nearest tenth of a second.

(ii) How many bacteria will there be after 6.8 seconds? Give your answer correct to the nearest hundred bacteria.

(iii) Will there be a time when there are no bacteria left? Explain your answer. *[6 marks]*

2. From November 2001 Paper 2 *[Maximum mark: 14]*

A rectangle has dimensions $(5 + 2x)$ metres and $(7 - 2x)$ metres.

(a) Show that the area, A, of the rectangle can be written as $A = 35 + 4x - 4x^2$. *[1 mark]*

(b) The following is the table of values for the function $A = 35 + 4x - 4x^2$.

x	−3	−2	−1	0	1	2	3	4
A	−13	p	27	35	q	r	11	s

(i) Calculate the values of p, q, r and s.

(ii) On graph paper, using a scale of 1 cm for 1 unit on the x-axis and 1 cm for 5 units on the A-axis, plot the points from your table and join them up to form a smooth curve. *[6 marks]*

(c) Answer the following, using your graph or otherwise.

(i) Write down the equation of the axis of symmetry of the curve.

(ii) Find one value of x for a rectangle whose area is 27 m^2.

(iii) Using this value of x, write down the dimensions of the rectangle. *[4 marks]*

(d) (i) On the same graph, draw the line with equation $A = 5x + 30$.

(ii) Hence or otherwise, solve the equation $4x^2 + x - 5 = 0$. *[3 marks]*

3. From May 2002 Paper 2 *[Maximum mark: 15]*

The graph below shows the tide heights, h metres, at time t hours after midnight, for *Tahini* island.

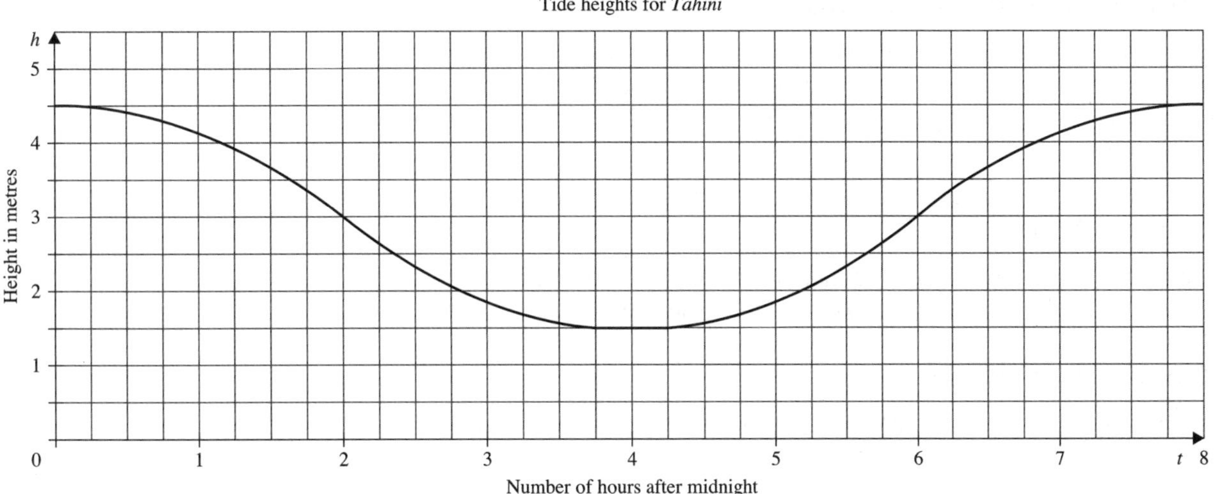

(a) Use the graph to find

(i) the height of the tide at 03:15;

(ii) the times when the height of the tide is 3.5 metres. *[3 marks]*

(b) The best time to catch fish is when the tide is **below** 3 metres. Find this best time, giving your answer as an inequality in t. *[3 marks]*

Due to the location of *Tahini* island, there is very little variation in the pattern of tidal heights. The maximum tide height is 4.5 metres and the minimum tide height is 1.5 metres. The height h can be modelled by the function

$$h(t) = a\cos(bt^{\circ}) + 3$$

(c) Use the graph above to find the values of the variables a and b. *[4 marks]*

(d) Hence **calculate** the height of the tide at 13:00. *[3 marks]*

(e) At what time would the tide be at its lowest point in the **second** 8-hour period? *[2 marks]*

TOPIC 19

Coordinates, midpoint and distance formulae

What do you need to know?

A graph has an x-axis and a y-axis, with points located using coordinates. You always need to label the x- and y-axes when doing sketches or graphs. In the IB Math Studies exam, for all graphs one mark is allocated for labelling the axes and creating a scale.

The midpoint of a line segment is the point that is the same distance away from both ends of the segment.

The distance formula is a way of calculating the distance between two points (or the length of a line segment if you know the coordinates of the endpoints).

How do you do it?

There are formulae for both distance and midpoint in your formula booklet:

5.1	Distance between two points (x_1, y_1) and (x_2, y_2)	$d = \sqrt{(x_1 - x_2)^2 + (y_1 - y_2)^2}$
	Coordinates of the midpoint of a line segment with endpoints (x_1, y_1) and (x_2, y_2)	$\left(\frac{x_1 + x_2}{2}, \frac{y_1 + y_2}{2}\right)$

So if you have two points, point 1 and point 2, you put the x-coordinate of the 1st point into x_1, the y-coordinate of the 1st point into y_1 and so on. It's the same idea for the midpoint formula – put the coordinates in the right places and calculate. Only two things to remember:

- The distance formula will give you a distance! As such, don't forget to write your answer to 3sf **with units** if they are given.
- The midpoint formula will give you the **coordinates** of the midpoint; make sure you write them as a set of coordinates.

Examples

1. Given the points A(3, 4) and B(–2, 8) the distance between A and B is $\sqrt{(3 - -2)^2 + (4 - 8)^2} = \sqrt{5^2 + (-4)^2} = \sqrt{25 + 16} = \sqrt{41} \approx 6.40$ (3sf).

   The coordinates of the midpoint of segment $\overline{AB}$ are $\left(\frac{3 + -2}{2}, \frac{4 + 8}{2}\right) = \left(\frac{1}{2}, 6\right)$.
2. There is a right-angled triangle ABC with coordinates A(–3, –2), B(–4, 1) and C(3, 0).

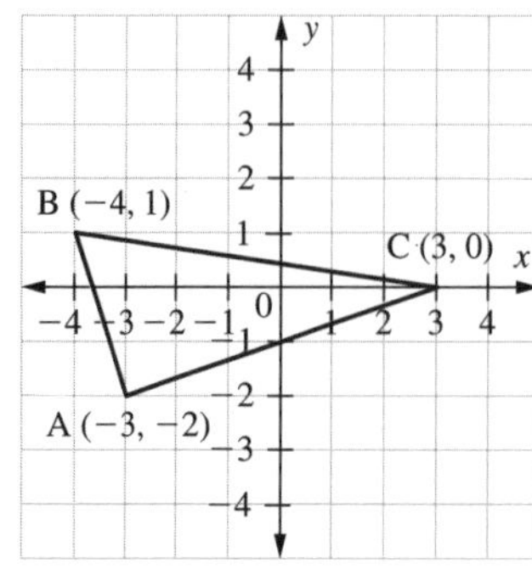

The lengths of the legs are

$AB = \sqrt{(-3 - -4)^2 + (-2 - 1)^2} = \sqrt{1^2 + (-3)^2} = \sqrt{1 + 9} = \sqrt{10}$ and

$AC = \sqrt{(-3 - 3)^2 + (-2 - 0)^2} = \sqrt{(-6)^2 + (-2)^2} = \sqrt{36 + 4} = \sqrt{40}$.

Therefore the area of the triangle is $A = \frac{1}{2} \cdot \sqrt{10} \cdot \sqrt{40} = \frac{1}{2}\sqrt{400} = 10$.

By the way, you can verify that the triangle is indeed right-angled by

- finding the length of BC and using Pythagoras' theorem:

  $BC = \sqrt{(-4 - 3)^2 + (1 - 0)^2} = \sqrt{50}$

  Does $AB^2 + AC^2 = BC^2$?

  $(\sqrt{10})^2 + (\sqrt{40})^2 = (\sqrt{50})^2 \Rightarrow 10 + 40 = 50$ ✓

  '(this is shown in the next section)'

- showing that the product of the gradients of lines AB and AC is –1:

  gradient of AB $= \frac{-2 - 1}{-3 - -4} = \frac{-3}{1} = -3$,

  gradient of AC $= \frac{0 - -2}{3 - -3} = \frac{2}{6} = \frac{1}{3} - 3 \cdot \frac{1}{3} = -1$ ✓

- showing that the angle is 90° by the cosine rule:

  $\cos\theta = \frac{(\sqrt{10})^2 + (\sqrt{40})^2 - (\sqrt{50})^2}{2 \cdot \sqrt{10} \cdot \sqrt{40}} = 0 \Rightarrow \theta = 90°$ ✓

  (again, this is shown in a later section)

How can my calculator help me do this?

There is no real way your GDC can help you here beyond just doing the arithmetic. Sorry!

Examples from past IB papers

An example from May 2000 Paper 2 (modified)

The line L_1 shown on the set of axes below cuts the x-axis at A (8, 0) and the y-axis at B (0, 6). The line L_2 goes through the midpoint of the line segment [AB] and the point C (0, –2).

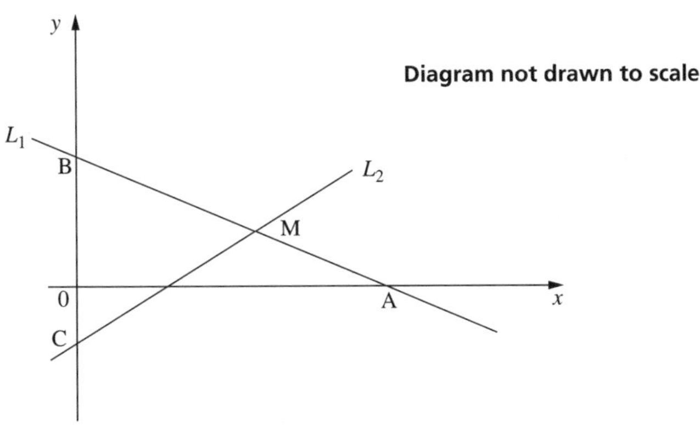

(a) Write down the coordinates of M, the midpoint of [AB].

(b) Find the length of

(i) MC;

(ii) AC.

Working:

(a) $M\left(\frac{(8+0)}{2}, \frac{(0+6)}{2}\right)$

$M(4, 3)$

(b) (i) $MC = \sqrt{(4-0)^2 + (3-(-2))^2}$

$= \sqrt{16 + 25}$

$= \sqrt{41} = 6.4031242$

(ii) $AC = \sqrt{(8-0)^2 + (0-(-2))^2}$

$= \sqrt{64 + 4}$

$= \sqrt{68} = 8.246\ 211\ 251$

Answers:

(a) (4, 3)

(b) (i) 6.40 (3sf)

(ii) 8.25 (3sf)

Now you practise it

An example from May 2004 Paper 2 (modified)

The coordinates of the vertices of a triangle are P (−2, 6), Q (6, 2) and R (−8, *a*), where $a < 0$.

(a) On the graph below, mark the points P and Q on a set of coordinate axes.

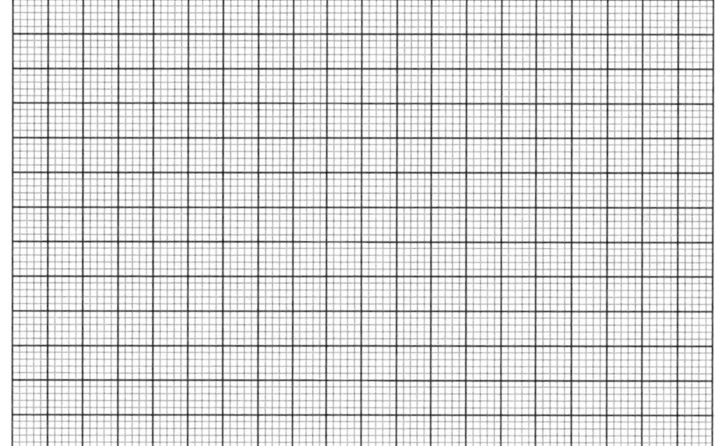

(b) Calculate the distance PQ.

(c) The length of PR is $\sqrt{180}$. Find *a*.

(d) Find the area of triangle PQR.

Working:

Answers:

(b) ....

(c) ....

(d) ....

TOPIC 20

Equation of a line: forms, slope/gradient, perpendicular and parallel lines

What do you need to know?

The slope, *m* (also called the gradient), of a line is a number telling you how steep the line is, or what direction it is going.

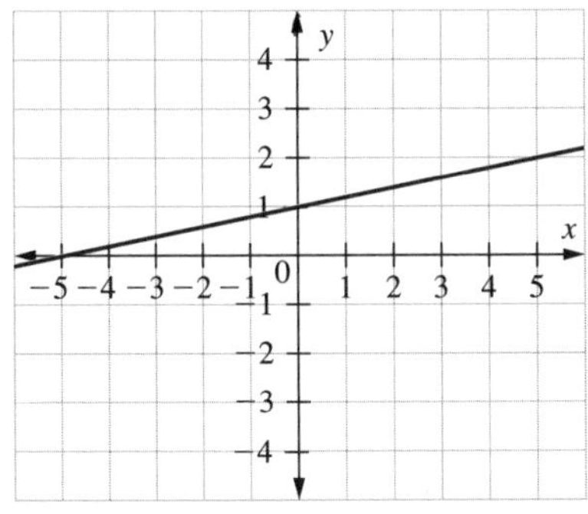

positive, small slope e.g. $m = 0.2$

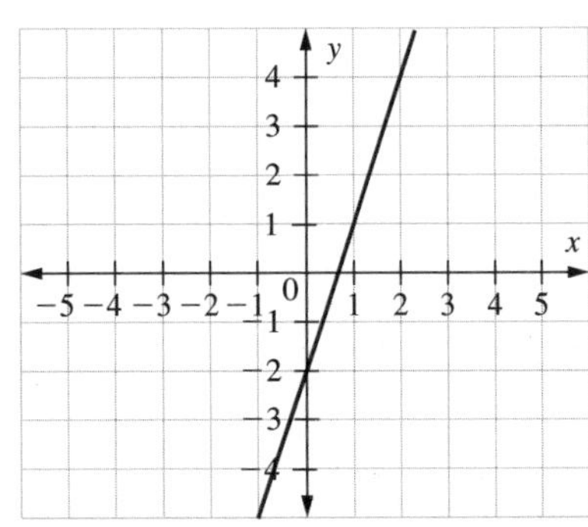

positive, large slope e.g. $m = 3$

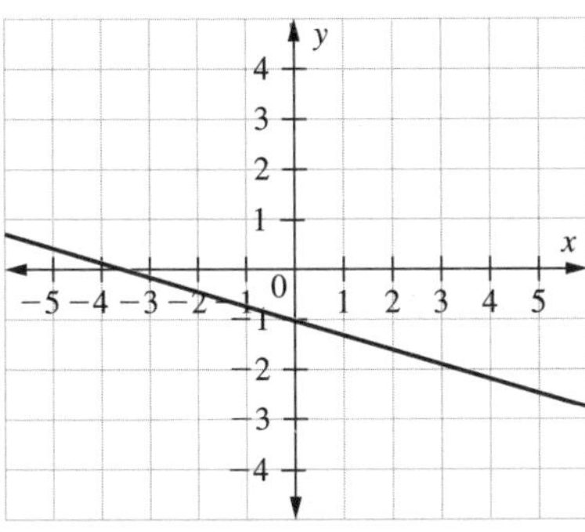

negative, small slope e.g. $m = -0.3$

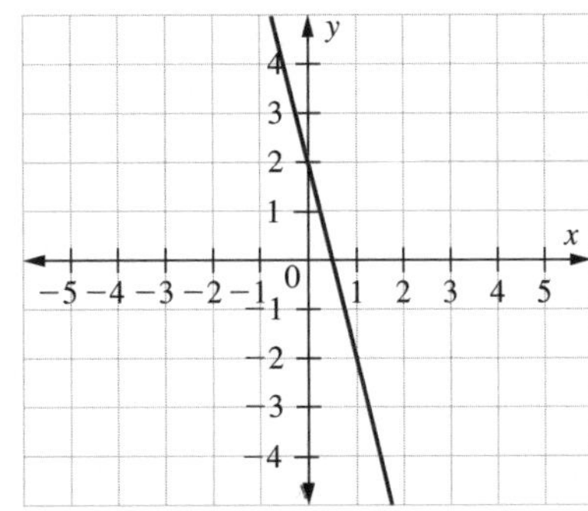

negative, large slope e.g. $m = -4$

Small slopes are between −1 and 1, large slopes are greater than 1 or less than −1, and lines that go up/down at 45° have a slope of exactly 1 or −1.

An equation of a line tells you how to calculate the coordinates of points anywhere on the line. This equation can be in several different forms:

- **Slope–intercept form**: $y = mx + c$, where x and y are the coordinates of points on the line, m is the slope of the line and c is the y-intercept of the line.
- **Standard form**: $ax + by + d = 0$ where again x and y are the coordinates of points on the line, $-\frac{d}{a}$ is the x-intercept and $-\frac{d}{b}$ is the y-intercept.

SCOTT SAYS:

I teach my students to play "peek-a-boo" with equations in standard form. If you want the x-intercept, cover the y term with your finger and solve what you see for x. Do the opposite for the y-intercept. If you lift your fingers up, you can say "peek-a-boo"!

- **Point–slope form**: $y - y_1 = m(x - x_1)$ where again x and y are the coordinates of points on the line, m is the slope and x_1 and y_1 are the coordinates of a point you already know is on the line.

Two lines that are parallel have exactly the same slope.

Two lines that are perpendicular have slopes that multiply together to make −1. They are "negative reciprocals" of each other, like $\frac{5}{3}$ and $-\frac{3}{5}$.

How do you do it?

To find the slope of a line, it depends on what you are given. If you are given

- the coordinates of two points, use the slope formula in your booklet:

5.2	Gradient formula	$m = \frac{y_2 - y_1}{x_2 - x_1}$

- the equation of the line in slope–intercept form, just write down m, the number in front of the x (or whatever variable is there)
- the equation of the line in standard form, the slope is $\frac{-a}{b}$.

To graph the equation of a line in slope–intercept form, plot the y-intercept and use the slope (e.g. a slope of $\frac{5}{3}$ would mean to go up 5 and 3 to the right) to plot a second point. Then draw a line through the points.

To graph the equation of a line in standard form, plot the x- and y-intercepts and draw a line through the points.

To find an equation of a line, use the slope formula to find the slope first. Then put the slope and a point into $y = mx + c$ to find c.

To convert an equation of a line from slope–intercept form to standard form, just move the x term and the y – intercept to the left-hand side.

To find a new line parallel to a given line, just use the same slope as the original line.

To find a new line perpendicular to a given line, use the negative reciprocal of the slope of the original line as the new slope (or use this trick: $\perp$ slope $= \frac{-1}{\text{old slope}}$).

Examples

1. Given the equation of a line $y = 3x + 6$, the slope is 3 and the y-intercept is 6.

   To graph it, place a point at (0, 6), and go up 3 and 1 to the right, and place a point there – you should be at (1, 9). Draw a line through (0, 6) and (1, 9).

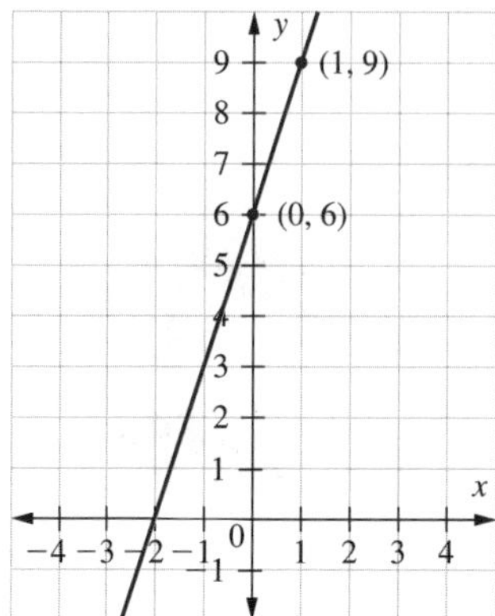

2. To find an equation of a line parallel to $y = 3x + 6$ that goes through (4, 5), start with a new equation with the same slope: $y = 3x + c$.

Put in (4, 5) and you get $5 = 3 \cdot 4 + c$.

Solving for c you get $c = -7$.

So the equation is $y = 3x - 7$.

In standard form this is $-3x + 7 = 0$.

Its x-intercept is $\frac{-7}{-3} = \frac{7}{3} = 2.33$ and its y-intercept is -7.

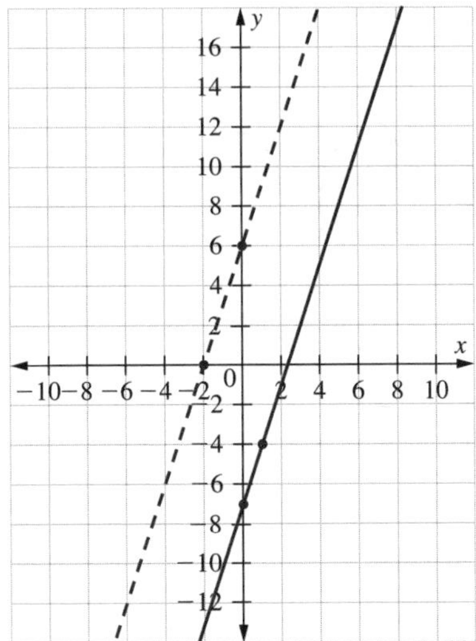

3. To find an equation of a line that goes through the origin and is perpendicular to a line that goes through (–1, 3) and (9, 4), first find the old slope:

$m = \frac{4-3}{9--1} = \frac{1}{10}$.

Now find the new slope: $\perp$ slope $= \frac{-1}{\frac{1}{10}} = -10$.

So the new equation is $y = -10x + c$.

As the line goes through the origin, the y-intercept is 0.

So the equation of the line is $y = -10x$.

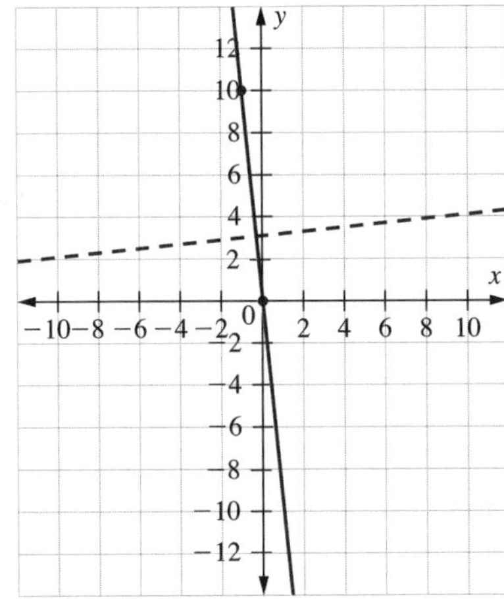

How can my calculator help me do this?

Your GDC is useful for graphing lines and finding x- and y-intercepts. Just be careful – you have to convert all equations to the form $y = mx + c$ in order to use the GDC.

Graphing $y = 3x + 6$

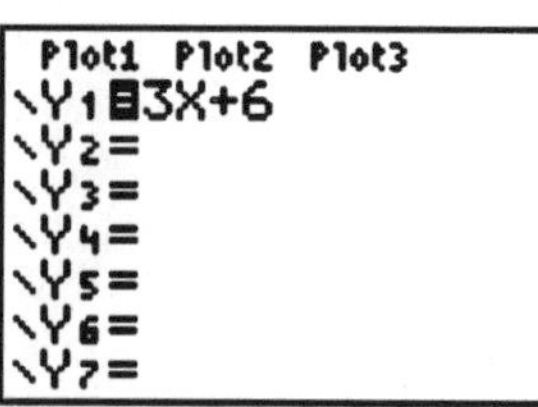

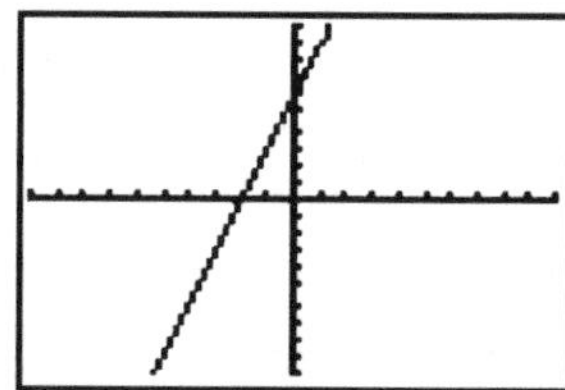

Finding the *x*- and *y*-intercepts of $2x - 7y - 11 = 0$ (Note that in order to enter the equation into Y=, you have to solve it for *y* first. Move the *x* term and the 11 over and then divide by the number next to the *y*.)

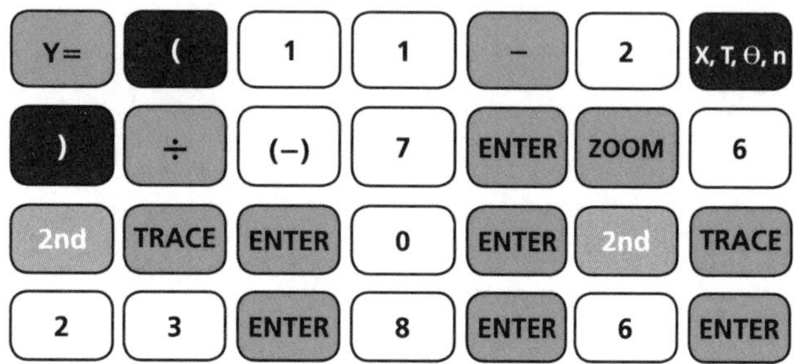

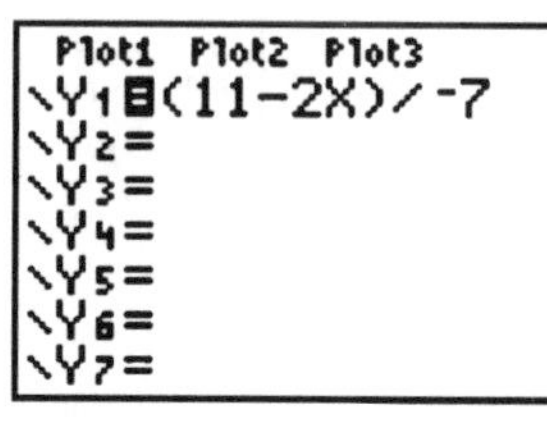

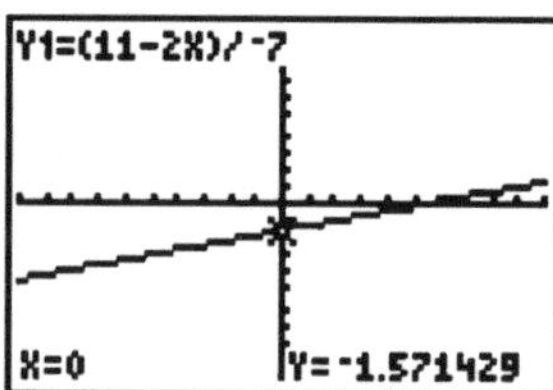

Finding the equation of the line that goes through (3, 4) and (–6, 7)

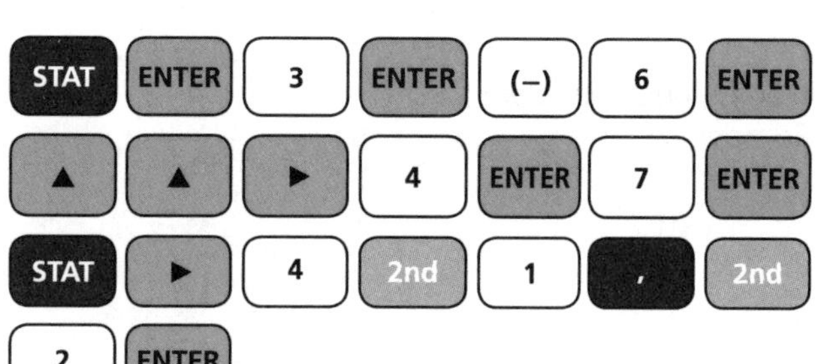

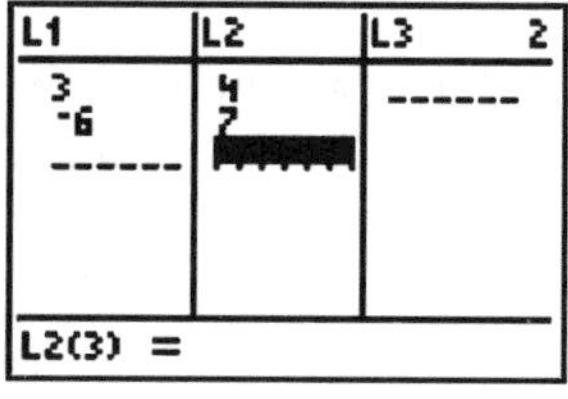

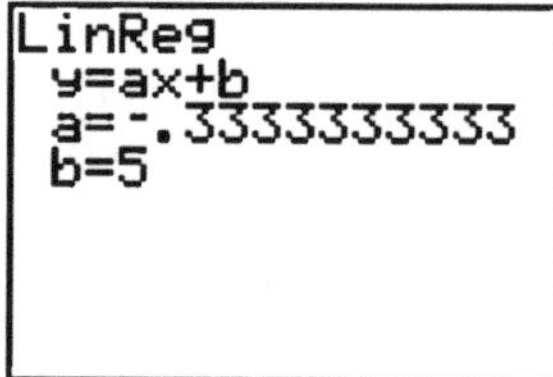

(Note that the answer is there but you have to write it in a different form:
$y = -\frac{1}{3}x + 5$ or $y = -0.333x + 5$)

Examples from past IB papers

An easy one from May 2002 Paper 1

The following diagram shows the lines l_1 and l_2, which are perpendicular to each other.

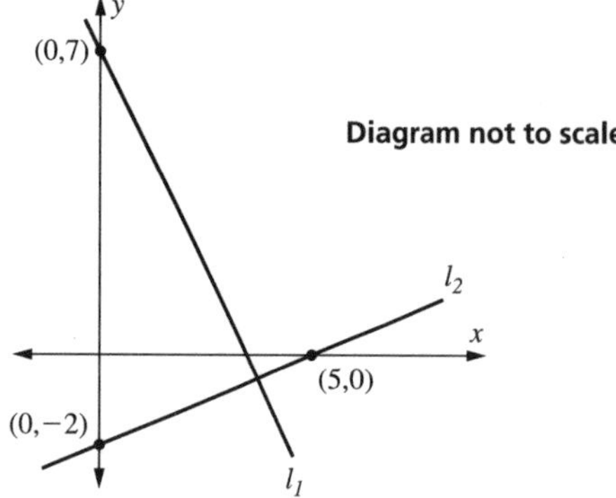

(a) Calculate the gradient of line l_1.

(b) Write the equation of line l_1 in the form $ax + by + d = 0$ where *a*, *b* and *d* are integers, and $a > 0$.

Working:

(a) gradient of $l_2 = \frac{(0 - (-2))}{(5 - 0)} = \frac{2}{5}$

gradient of $l_1 = -\frac{5}{2}$

(b) $y = mx + c$

$7 = -\frac{5}{2}(0) + c$

$7 = c$

$y = -\frac{5}{2}x + 7$

$2y = -5x + 14$

$5x + 2y - 14 = 0$

Answers:

(a) $-\frac{5}{2}$

(b) $5x + 2y - 14 = 0$

A harder one from November 2007 Paper 1

The mid-point, M, of the line joining A(s, 8) to B(−2, t) has coordinates M(2, 3).

(a) Calculate the values of s and t.

(b) Find the equation of the straight line perpendicular to AB, passing through the point M.

Working:

(a) midpoint formula

$\left(\frac{(s + (-2))}{2}, \frac{(8 + t)}{2}\right) = (2, 3)$

$\frac{(s - 2)}{2} = 2, \frac{(8 + t)}{2} = 3$

$s = 6, t = -2$

(b) gradient of AB $= \frac{(8 - (-2))}{(6 - (-2))} = \frac{10}{8} = \frac{5}{4}$

perpendicular: $m = -\frac{4}{5}$

$y = -\frac{4}{5}x + c$

$3 = -\frac{4}{5}(2) + c$

$\frac{23}{5} = c$

$y = -\frac{4}{5}x + \frac{23}{5}$

Answers:

(a) $s = 6, t = -2$

(b) $y = \frac{-4}{5}x + \frac{23}{5}$

Now you practise it

An easy one from November 2001 Paper 1

The equation of a line l_1 is $y = \frac{1}{2}x$.

(a) On the grid, draw and label the line l_1.

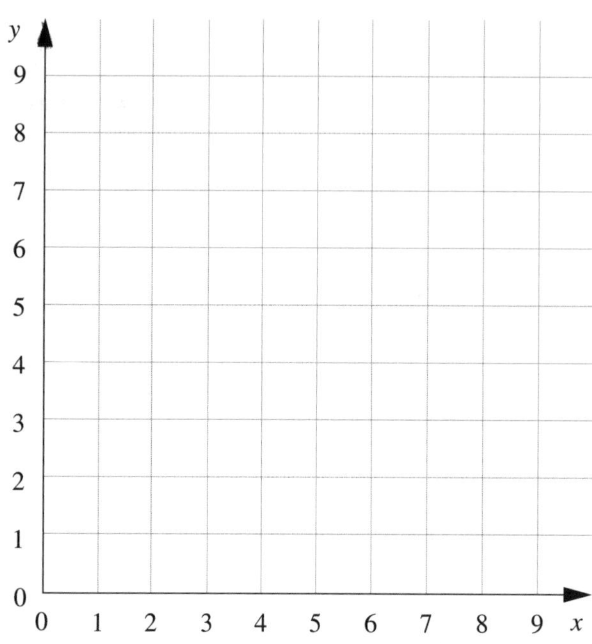

The line l_2 has the same gradient as l_1 but crosses the y-axis at 3.

(b) What is the geometric relationship between l_1 and l_2?

(c) Write down the equation of l_2.

(d) On the same grid as in part (a), draw the line l_2.

Working:

Answers:

(b) ....

(c) ....

A harder one from May 2001 Paper 1

A is the point (2, 3), and B is the point (4, 9).

(a) Find the gradient of the line segment [AB].

(b) Find the gradient of a line perpendicular to the line segment [AB].

(c) The line $2x + by - 12 = 0$ is perpendicular to the line segment [AB]. What is the value of b?

Working:

Answers:

(a) ....

(b) ....

(c) ....

TOPIC 21

Right-angled triangle trigonometry (SOHCATOA)

What do you need to know?

Right-angled triangle trig (RTT) is used to find the sizes of angles or the lengths of sides of right-angled triangles. You use three formulae:

$$\sin \theta = \frac{\text{opposite}}{\text{hypotenuse}} \text{ (SOH)}$$

$$\cos \theta = \frac{\text{adjacent}}{\text{hypotenuse}} \text{ (CAH)}$$

$$\tan \theta = \frac{\text{opposite}}{\text{adjacent}} \text{ (TOA)}$$

hypotenuse
opposite
θ
adjacent

How do you do it?

Step 1: Make sure that you are indeed able to use RTT – you must have a right-angled triangle with at least two known sides or a known side and a known angle.

Step 2: Identify the corresponding "opposite", "adjacent" and "hypotenuse" sides (the "adjacent" is the side next to the angle, the "hypotenuse" is the longest side opposite the right angle, and the "opposite" is the side opposite the angle).

Step 3: Use the appropriate formula from above and solve for the missing side or angle.

HANA AND CHRIS SAY:

If you know two sides of a right-angled triangle and are missing the third side, *do not* use RTT. Use Pythagoras' theorem: $a^2 + b^2 = c^2$!

Examples

1.

13 cm
62°
x cm

$$\sin 62° = \frac{13}{x}$$

$$x = \frac{13}{\sin 62°}$$

$$x = 14.723\,410\,66$$

$$x = 14.7 \text{ cm (3s.f.)}$$

2.

θ
1.1 mm
1.7 mm

$$\tan \theta = \frac{1.7}{1.1}$$

$$\theta = \tan^{-1}\left(\frac{1.7}{1.1}\right)$$

$$\theta = 57.094\,757\,08$$

$$\theta = 57.1° \text{ (3s.f.)}$$

How can my calculator help me do this?

You use your calculator to find the sin, cos or tan of various ratios, or use the $\sin^{-1}$, $\cos^{-1}$ or $\tan^{-1}$ functions to find angles.

Make sure your calculator is set to degrees (again remember that when your invigilator resets your calculator before the exam, you need to do this!):

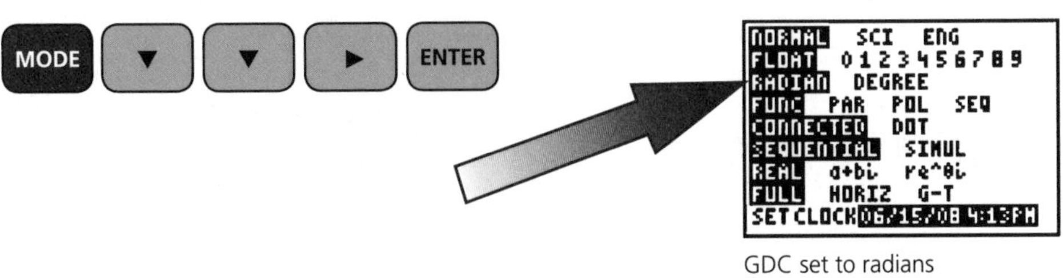

GDC set to radians
(default – BAD)

GDC set to degrees
(not default – GOOD)

Finding the missing side in the first example:

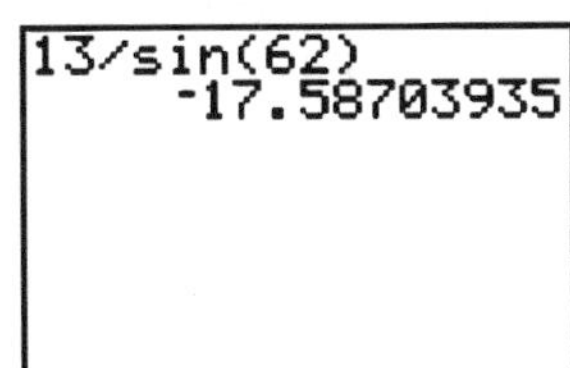

Oops – GDC was set to radians. Notice that the answer makes no sense.

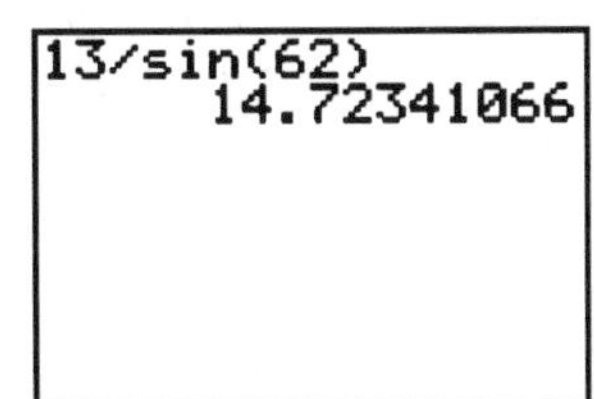

Ah – this makes a lot more sense! Don't forget to round to 3sf afterwards.

Finding the missing angle in the second example:

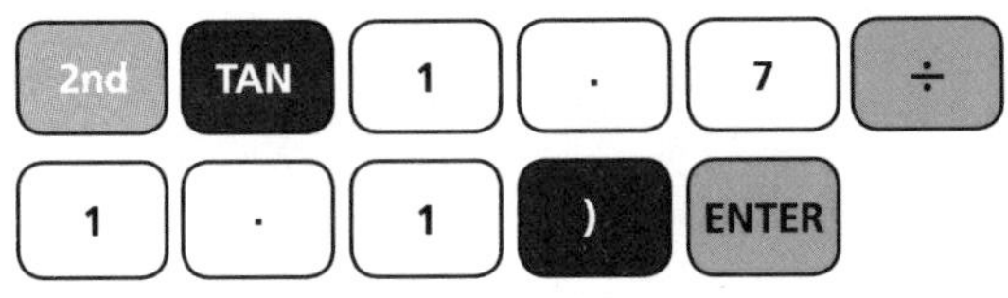

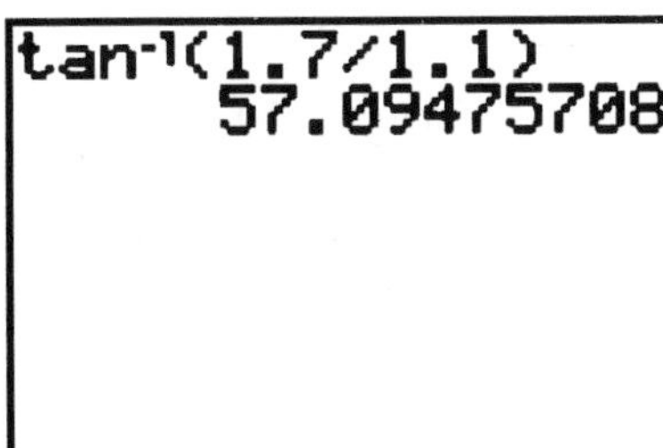

Examples from past IB papers

An easy one from May 2002 Paper 1

Andrew is at point A in a park. A deer is 3 km directly north of Andrew, at point D. Brian is 1.8 km due west of Andrew, at point B.

(a) Draw a diagram to represent this information.

(b) Calculate the distance between Brian and the deer.

(c) Brian looks at Andrew, and then turns through an angle θ to look at the deer. Calculate the value of θ.

Working:

(a)

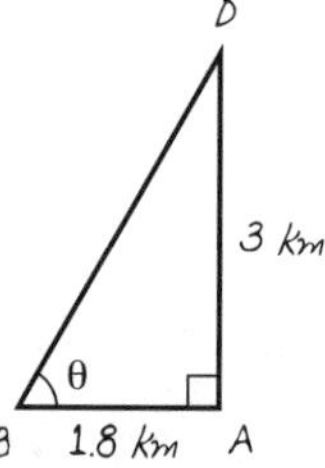

(b) $BD = \sqrt{(3^2 + 1.8^2)} = \sqrt{12.24} = 3.498571$

(c) $\tan \vartheta = \frac{3}{1.8}$

$\vartheta = 59.036243°$

Answers:

(a) see diagram

(b) 3.50 km (3sf)

(c) 59.0° (3sf)

A harder one from November 2004 Paper 1

OABCD is a square-based pyramid of side 4 cm as shown in the diagram. The vertex D is 3 cm directly above X, the centre of square OABC. M is the mid-point of AB.

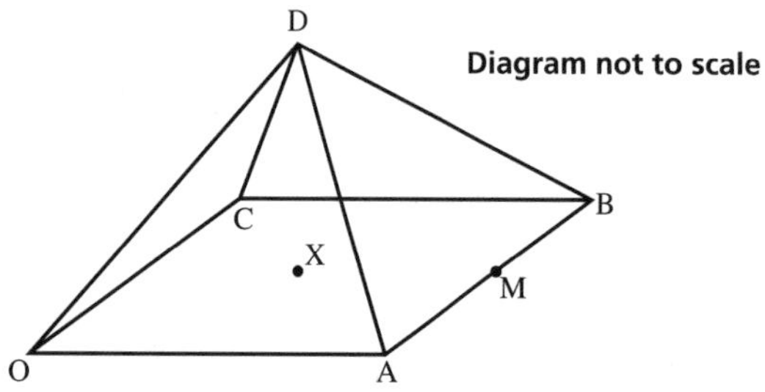

(a) Find the length of XM.

(b) Calculate the length of DM.

(c) Calculate the angle between the face ABD and the base OABC.

Working:

(b)

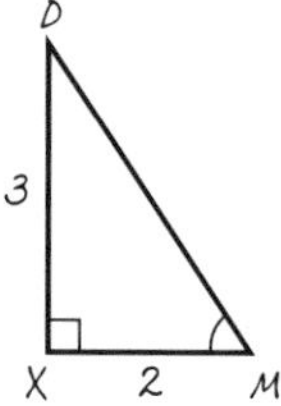

$DM = \sqrt{(3^2 + 2^2)}$

$= \sqrt{13}$ or 3.605551

(c) $\tan DMX = \frac{3}{2}$

angle $DMX = 56.309932°$

Answers:

(a) XM = 2 cm

(b) DM = 3.61 (3sf)

(c) DMX = 56.3° (3sf)

Now you practise it

An easy one from November 2001 Paper 1

The diagram shows a cuboid 22.5 cm by 40 cm by 30 cm.

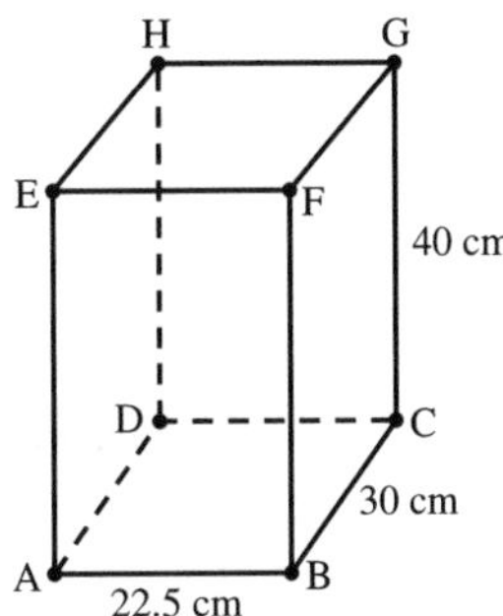

(a) Calculate the length of [AC].

(b) Calculate the size of $G\hat{A}C$.

Working:

Answers:

(a) ....

(b) ....

A harder one from May 2001 Paper 1

A rectangular block of wood with face ABCD leans against a vertical wall, as shown in the diagram below. AB = 8 cm, BC = 5 cm and angle BAE = 28°.

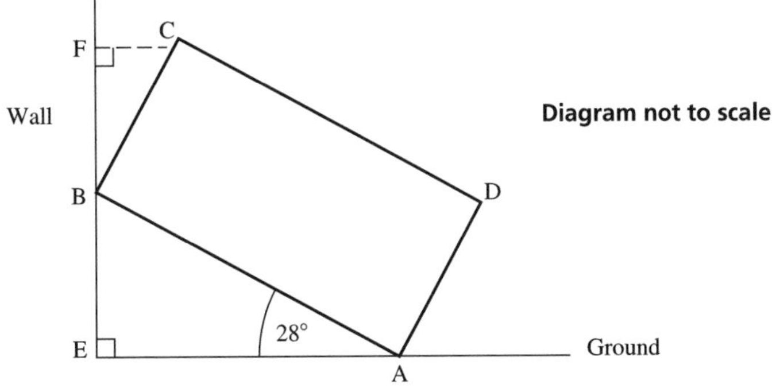

Find the vertical height of C above the ground.

Working:

Answers:

....

TOPIC 22

Sine rule and area of a triangle

What do you need to know?

The sine rule (Law of Sines) is used to find the measures of angles or lengths of sides of non-right-angled triangles where you already know a "pair" of information: an angle and a side opposite each other.

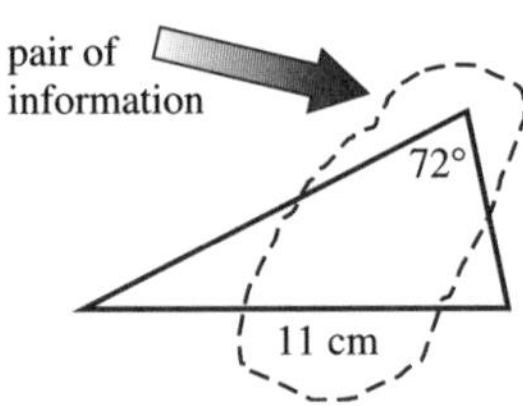

The formula for the sine rule is in your booklet:

5.4	Sine rule	$\frac{a}{\sin A} = \frac{b}{\sin B} = \frac{c}{\sin C}$

a, b and c are the lengths of the sides opposite vertices A, B and C on the triangle.

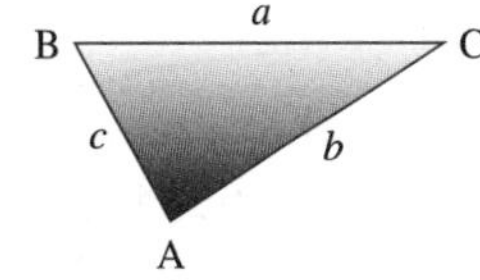

The area of a triangle is normally found by the formula $A = \frac{1}{2}bh$ but there is another one using trigonometry in your formula booklet:

5.4	Area of a triangle	Area of a triangle $= \frac{1}{2}ab \sin C$, where a and b are adjacent sides, C is the included angle

How do you do it?

When you use the sine rule, you only use two of the three fractions at a time. It is irrelevant how you name the sides and angles as long as you match up the A with the a, the B with the b and the C with the c. Once you put the information into the formula, *cross-multiply* the two fractions to get an equation and then solve. Make sure that your "pair" is one of the two fractions you're using.

To find the area of a triangle using the formula in 5.4, just substitute the numbers. This is an easy one. But make sure that the angle you're using is in between the two known sides.

SCOTT SAYS:

I tell my students that in order to find the area, you have to make a sandwich: each side is a piece of bread and the angle is inside. If the angle is not inside the bread, you cannot use $\frac{1}{2}ab \sin C$.

Examples

1.

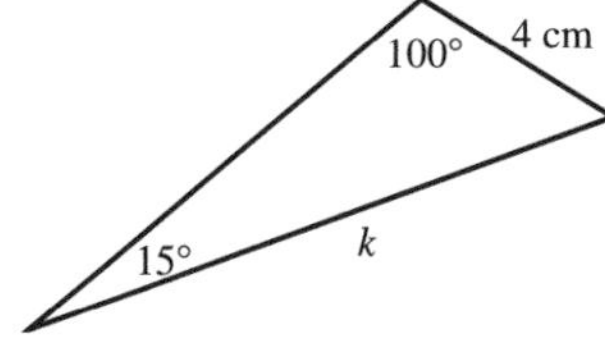

$$\frac{4}{\sin 15°} = \frac{k}{\sin 100°}$$

$$4 \sin 100° = k \sin 15°$$

$$k = \frac{4 \sin 100°}{\sin 15°}$$

$$k = 15.220\ 019\ 88$$

$$k = 15.2 \text{ cm (3sf)}$$

$$\text{Area} = \frac{1}{2} \cdot 4 \cdot 15.2 \cdot \sin 65°$$
$$= 27.6 \text{ cm}^2 \text{ (3sf)}$$

2.

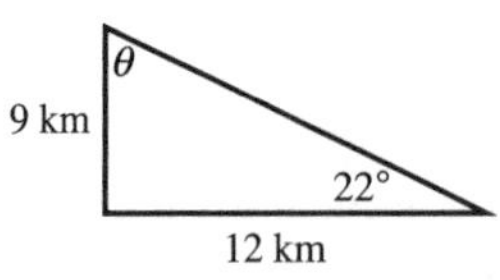

$$\frac{9}{\sin 22°} = \frac{12}{\sin \theta}$$

$$9 \sin \theta = 12 \sin 22°$$

$$\sin \theta = \frac{12 \sin 22°}{9}$$

$$\sin \theta = 0.499\ 475\ 457\ 9$$

$$\theta = \sin^{-1}(0.499\ 475\ 457\ 9)$$

$$\theta = 29.965\ 302\ 64°$$

$$\theta = 30.0° \text{ (3sf)}$$

$$\text{Area} = \frac{1}{2} \cdot 9 \cdot 12 \cdot \sin 128.0°$$
$$= 42.6 \text{ km}^2 \text{ (3sf)}$$

IRENE SAYS:

Did you notice that I drew the second triangle so that it looked like it was a right-angled triangle? I do this all the time. Be careful! If I don't indicate the triangle is right-angled, either by saying that the angle is 90° or by drawing a little □ in the corner, it's not necessarily a right-angled triangle.

How can my calculator help me do this?

Again, like with SOHCAHTOA, the calculator is only useful here in finding the values of sin or $\sin^{-1}$. Nothing new here.

Examples from past IB papers

An easy one from the November 1999 Paper 1 (modified)

The diagram shows the plan of a playground with dimensions as shown.

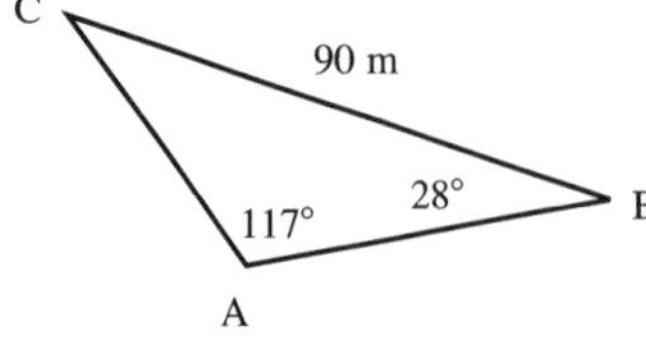

Calculate

(a) the length AC;

(b) the area of triangle ABC.

Working:

(a) $\frac{AC}{\sin 28} = \frac{90}{\sin 117}$

$AC = 47.42102$

(b) $Area = (\frac{1}{2})ab \sin C$

angle $C = 180 - (117 + 28) = 35°$

$Area = (\frac{1}{2} \times 90 \times 47.4) \sin 35$

$= 1223.4385$

Answers:

(a) 47.4 m (3sf)

(b) 1220 m^2 (3sf)

A harder one from November 2006 Paper 1

The figure shows a triangular area in a park surrounded by the paths AB, BC and CA, where AB = 400 m, $A\hat{B}C = 50°$ and $B\hat{C}A = 30°$.

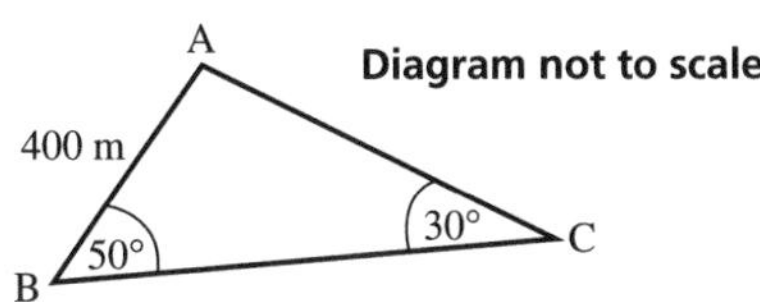

(a) Find the length of AC using the above information.

Diana goes along these three paths in the park at an average speed of 1.8 ms^{-1}.

(b) Given that BC = 788 m, calculate how many minutes she takes to walk once around the park.

Working:

(a) $\frac{AC}{\sin 50} = \frac{400}{\sin 30}$

$AC \sin 30 = 400 \sin 50$

$AC = 612.83555$

(b) Distance $= 400 + 613 + 788$

$= 1801$ m

Time $= \frac{1801}{1.8} = 1000.555...$ seconds

$= 16.6759$ minutes

Answers:

(a) 613 m (3sf)

(b) 16.7 min (3sf)

Now you practise it

An easy one from May 2001 Paper 1

The following diagram shows a triangle ABC. AB = 8 m, AC = 14 m, BC = 18 m, and $B\hat{A}C = 110°$.

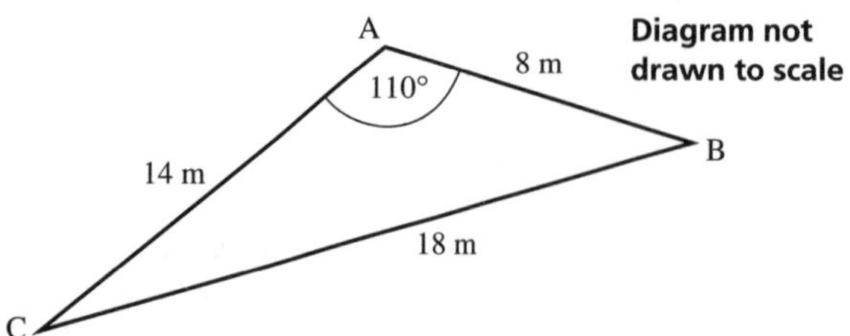

Calculate:

(a) the area of triangle ABC;

(b) the size of angle $A\hat{C}B$.

Working:

Answers:

(a) ....

(b) ....

A harder one from May 1999 Paper 1

In the diagram, triangle ABC is isosceles. AB = AC, CB = 15 cm and angle ACB is 23°.

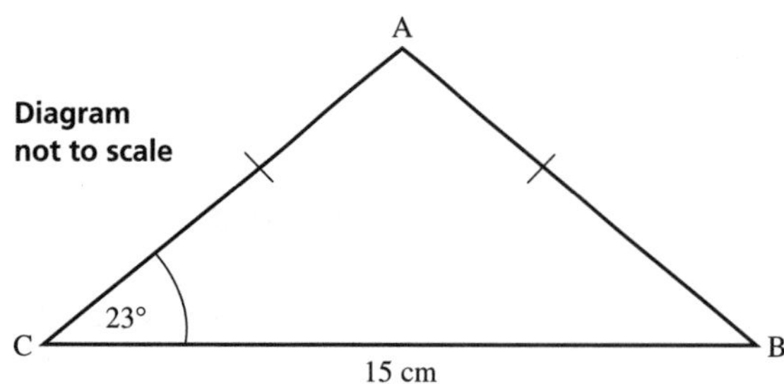

Find

(a) the size of angle CAB;

(b) the length of AB.

Working:

Answers:

(a) ....

(b) ....

TOPIC 23

Cosine rule

What do you need to know?

The cosine rule (Law of Cosines) is used to find the measures of angles or lengths of sides of non-right-angled triangles where you *do not* know a "pair" of information. Instead, you usually know either all three sides or two sides and an angle.

The formula for the cosine rule is in your booklet:

5.4	Cosine rule	$a^2 = b^2 + c^2 - 2bc\cos A; \quad \cos A = \dfrac{b^2 + c^2 - a^2}{2bc}$

a, b and c are the lengths of the sides opposite vertices A, B and C on the triangle.

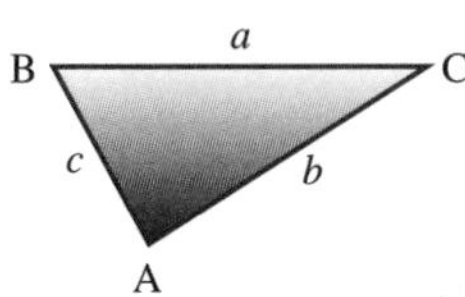

Sometimes students find it helpful to follow this chart in order to see which formula to use when doing triangle trig:

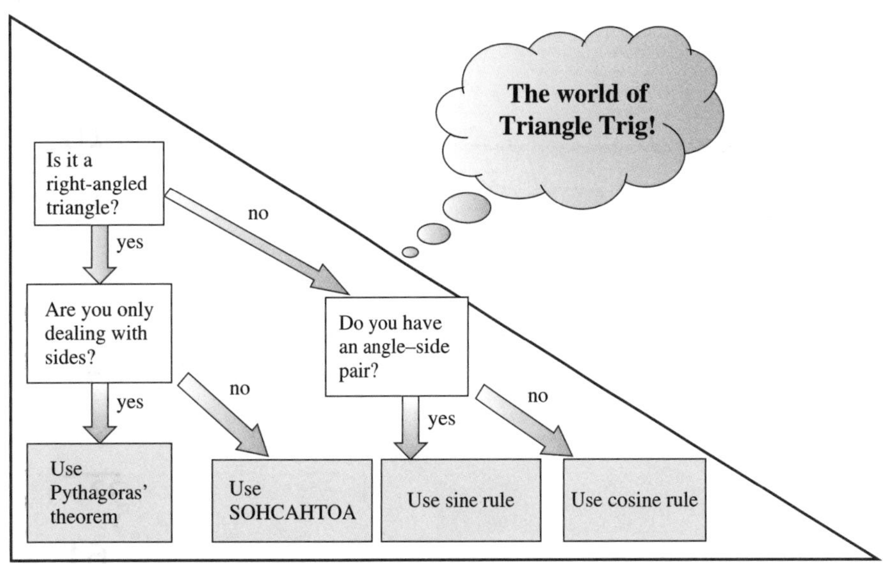

How do you do it?

The two formulae in the booklet are actually the same formula – one version is used when you are looking for a side (the first one) and the other version is used when you are looking for an angle (the second one). Just put in the information you have (sides are the small letters, angles are the large letters) and solve. Easy!

SCOTT SAYS:

Watch out! The most common mistake students make when doing the cosine rule is to forget the order of operations. Don't do the subtraction before multiplying the $2bc\cos A$!

Examples

1.

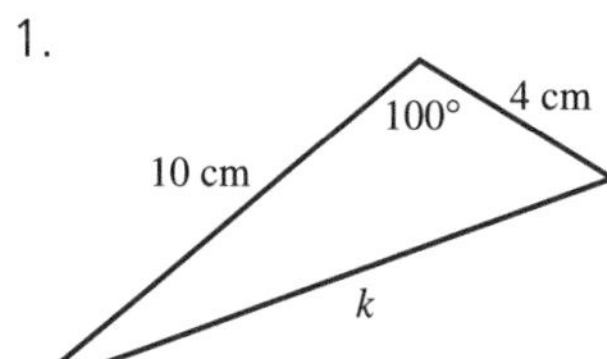

$k^2 = 10^2 + 4^2 - 2 \cdot 10 \cdot 4 \cdot \cos 100°$

$k^2 = 100 + 16 - 80 \cos 100°$

$k^2 = 116 - (-13.89185)$

$k^2 = 129.89185$

$k = 11.4$ cm (3sf)

2.

θ
9 km
12 km
11 km

$\cos\theta = \dfrac{12^2 + 9^2 - 11^2}{2 \cdot 12 \cdot 9}$

$\cos\theta = \dfrac{104}{216}$

$\theta = \cos^{-1}(0.48148)$

$\theta = 61.2°$ (3sf)

How can my calculator help me do this?

There isn't too much new material here for the GDC except for the fact that you can type the whole formula into your GDC at once. There is no need to manipulate or simplify the formula – just write the formula with the correct numbers on the exam and then write the answer!

How to do the first example…

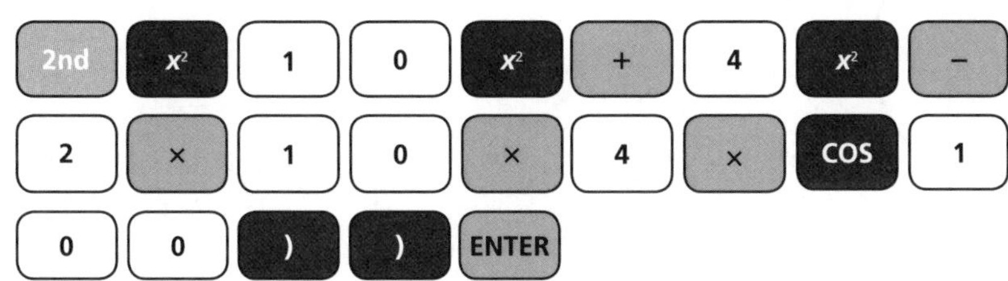

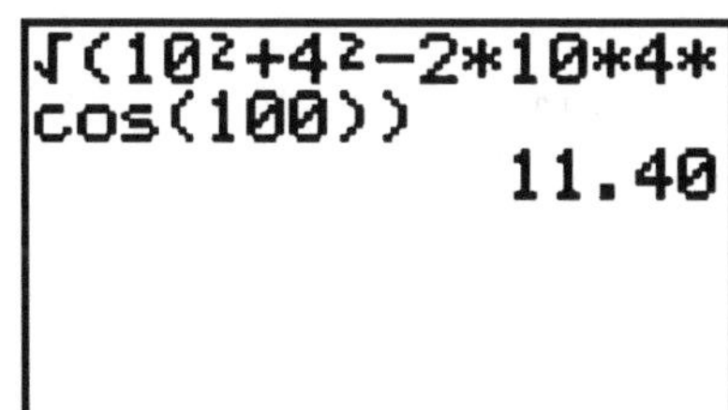

How to do the second example…

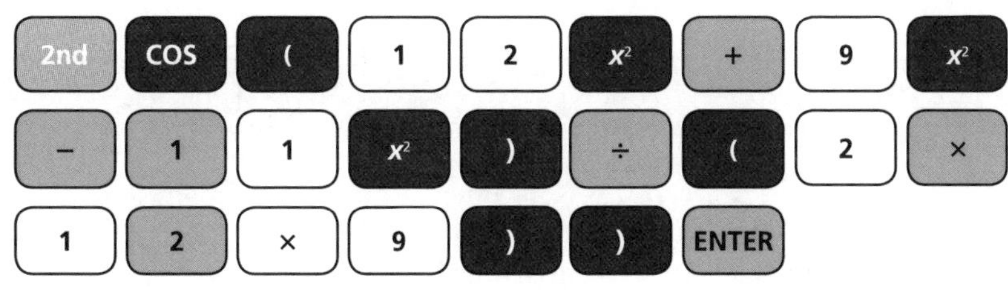

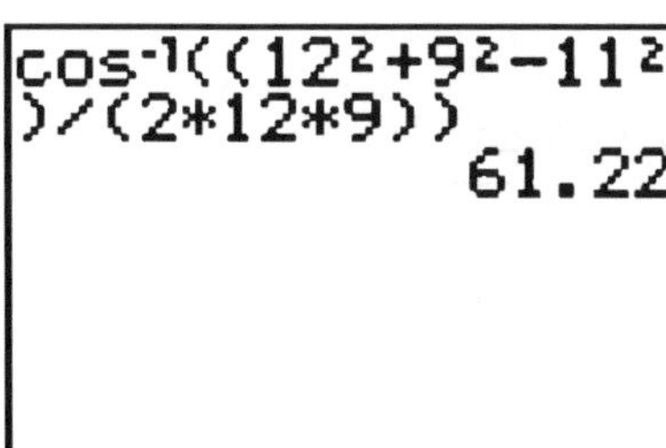

Examples from past IB papers

An easy one from November 2000 Paper 1

The diagram below shows an equilateral triangle ABC, with each side 3 cm long. The side [BC] is extended to D so that CD = 4 cm.

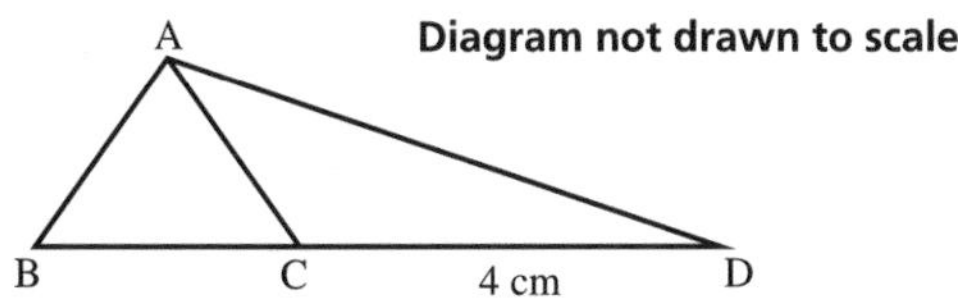

Calculate, **correct to two decimal places**, the length of [AD].

Working:

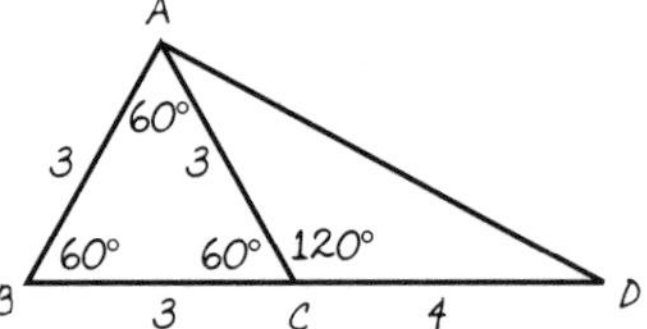

Angle ACD = 120°

$AD^2 = 3^2 + 4^2 - 2(3 \times 4) \cos 120$

$= 9 + 16 - 24 \cos 120 = 37$

$AD = \sqrt{37} = 6.0827625$

Answer:

6.08 cm (3sf)

A harder one from May 2006 Paper 1

The diagram shows a circle of radius R and centre O. A triangle AOB is drawn inside the circle. The vertices of the triangle are at the centre, O, and at two points A and B on the circumference. Angle AOB is 110 degrees.

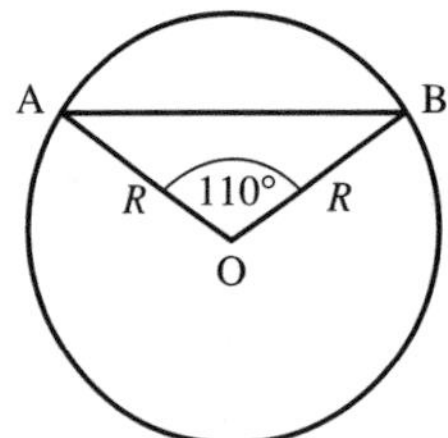

(a) Given that the area of the circle is 36π cm^2, calculate the length of the radius R.

(b) Calculate the length AB.

(c) Write down the side length L of a square which has the same area as the given circle.

Working:

(a) Area $= \pi R^2 = 36\pi$

$R = 6$ cm

(b) $AB^2 = 6^2 + 6^2 - 2(6 \times 6) \cos 110$

$= 72 - 72 \cos 110$

$= 96.62545$

$AB = 9.82982$

(c) $L = \sqrt{36\pi} = 6\sqrt{\pi} = 10.63472$

Answers:

(a) 6 cm

(b) 9.83 cm (3sf)

(c) 10.6 cm (3sf)

Now you practise it

An easy one from November 2007 Paper 1

On a map three schools A, B and C are situated as shown in the diagram.

Schools A and B are 625 metres apart.
Angle ABC = 102° and BC = 986 metres.

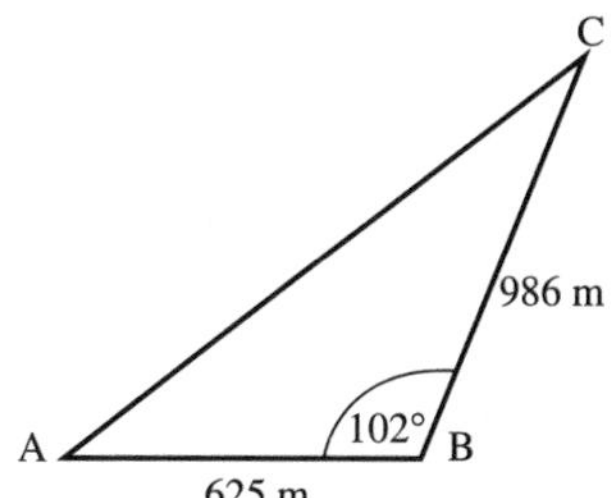

(a) Find the distance between A and C.

(b) Find the size of angle BAC.

Working:

Answers:

(a) ....

(b) ....

A harder one from May 2003 Paper 1

The figure below shows a hexagon with sides all of length 4 cm and with centre at O. The interior angles of the hexagon are all equal.

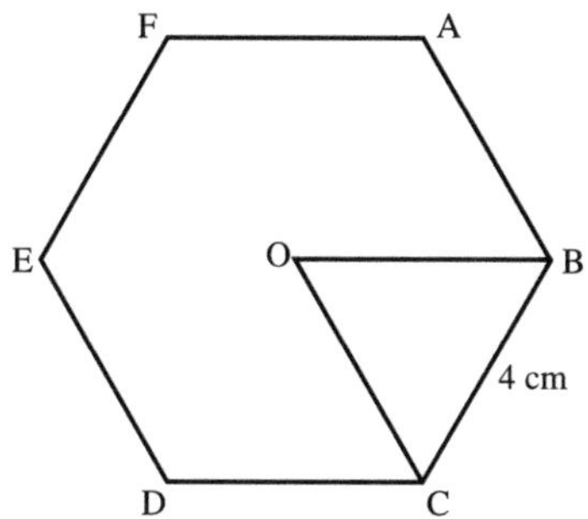

The interior angles of a polygon with n equal sides and n equal angles (regular polygon) add up to $(n - 2) \times 180°$.

(a) Calculate the size of angle $A\hat{B}C$.

(b) Given that OB = OC, find the area of the triangle OBC.

(c) Find the area of the whole hexagon.

Working:

Answers:

(a) ....

(b) ....

(c) ....

OPIC 24

Geometry of 3-D shapes

What do you need to know?

It is required to recognise and use the formulae for finding the surface area (SA) and volume (V) of various 3-dimensional objects:

cuboid

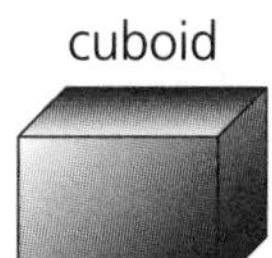

SA of cuboid = sum of areas of each rectangle
V of cuboid = length · width · height

prism

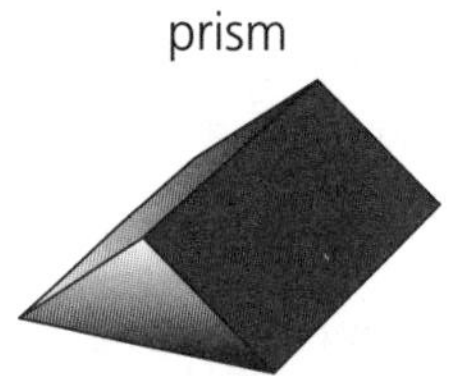

SA of prism = 2 · area of base + sum of areas of rectangles
V of prism = area of base · height

pyramid

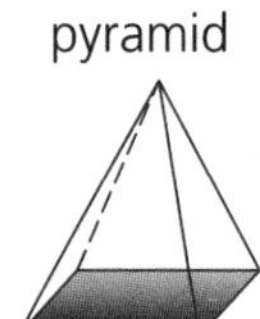

SA of pyramid = area of base + sum of areas of triangles
V of pyramid = $\frac{1}{3}$ · area of base · height

cylinder

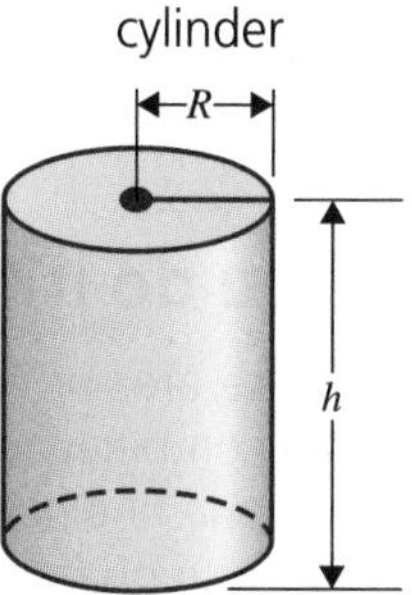

SA of cylinder = $2\pi R^2 + 2\pi Rh$ where R is the radius and h is the height
V of cylinder = $\pi R^2 h$

sphere

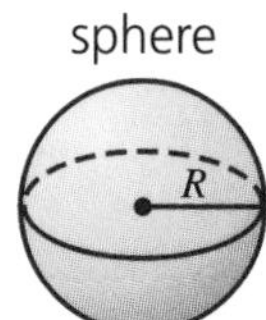

SA of sphere = $4\pi R^2$ where R is the radius
V of sphere = $\frac{4}{3}\pi R^3$

hemisphere

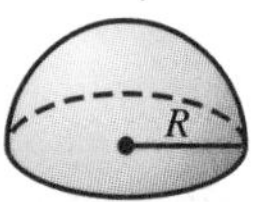

SA of hemisphere = $2\pi R^2$
V of hemisphere = $\frac{2}{3}\pi R^3$

cone

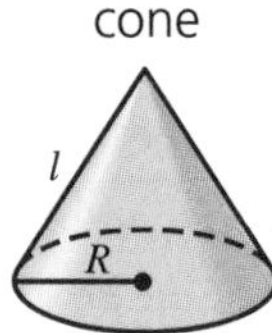

SA of cone = $\pi Rl + \pi R^2$ where R is the radius and l is the slant height
V of cone = $\frac{1}{3}\pi R^2 h$

Some, but not all, of these formulae are in your booklet in slightly different forms:

5.5	Volume of a pyramid	$V = \frac{1}{3}$(area of base × vertical height)
	Volume of a cuboid	$V = l \times w \times h$, where l is the length, w is the width, h is the height
	Volume of a cylinder	$V = \pi r^2 h$, where r is the radius, h is the height
	Area of the curved surface of a cylinder	$A = 2\pi rh$, where r is the radius, h is the height
	Volume of a sphere	$V = \frac{4}{3}\pi r^3$, where r is the radius
	Surface area of a sphere	$A = 4\pi r^2$, where r is the radius
	Volume of a cone	$V = \frac{1}{3}\pi r^2 h$, where r is the radius, h is the height
	Area of the curved surface of a cone	πrl, where r is the radius, l is the slant height

You also need to know how to calculate angles and lengths *inside prisms* (including cuboids) *and square-based pyramids only.*

How do you do it?

To find the surface area or volume of a 3-dimensional figure, just use the formula. If the figure is a combination of two or more objects, just use them one at a time (e.g. ice cream cone = a hemisphere + a cone!). Don't forget to write your units when you find these quantities – unit² for surface area and unit³ for volume. There's a one-mark penalty per paper if you forget your units!

To find an angle or a length inside a prism or pyramid, draw a triangle that has the information and then use triangle trigonometry to solve, as shown earlier in this section.

Examples

1.

$SA = 2 \cdot 17 \cdot 3 + 2 \cdot 17 \cdot 8 + 2 \cdot 3 \cdot 8$

$= 102 + 272 + 48$

$= 422 \text{ mm}^2$

To find AB, first find lengths of the sides of the triangle (indicated by dashed lines). Note that triangle ABC is right-angled: ∠ ACB = 90°.

$AC = \sqrt{17^2 + 3^2} = \sqrt{289 + 9} = \sqrt{298} \approx 17.26268$

$BC = 8$

So

$AB = \sqrt{17.26268^2 + 8^2} = \sqrt{298 + 64}$

$= \sqrt{362} \approx 19.0 \text{ mm (3sf)}$

2.

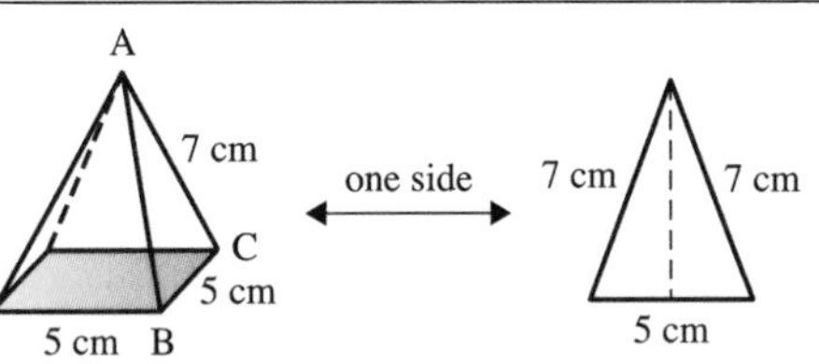

Height of one side (dashed line) =

$\sqrt{7^2 - 2.5^2} = \sqrt{49 - 6.25} = \sqrt{42.75} = 6.5383484$

Area of one side $= \frac{1}{2} \cdot 5 \cdot 6.5383484 = 16.345871$

$SA = 4 \cdot \text{area of one side} + 5 \cdot 5$

$= 65.383484 + 25$

$= 90.4 \text{ cm}^2 \text{ (3sf)}$

Angle CBA is found easiest by SOHCAHTOA:

$\cos\theta = \frac{2.5}{7} = 0.35714$

$\theta = \cos^{-1}(0.35714) \approx 69.1° \text{ (3sf)}$

How can my calculator help me do this?

Your GDC is not helpful here at all!

Examples from past IB papers

A question in the style of Paper 1

An ice cream dessert consists of an ice cream-filled wafer cone with a hemisphere of extra ice cream on top. The dimensions of the dessert are shown below.

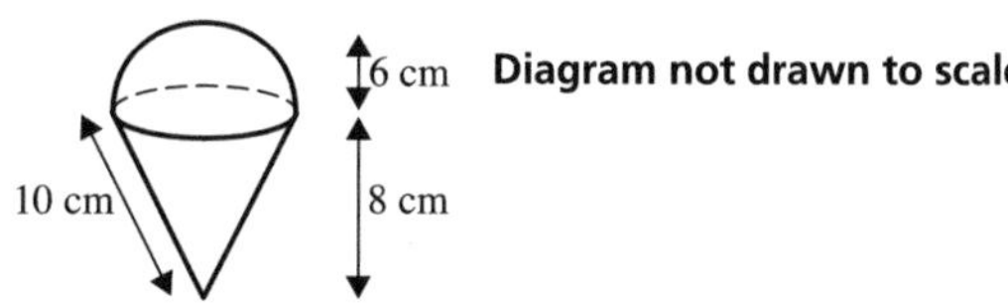

Diagram not drawn to scale

(a) If the wafer cone is covered with a paper wrapper, find the surface area of the paper.

(b) Find the total volume of the ice cream.

Working:

(a) $A = \pi r l$

$= \pi(6)(10)$

$= 60\pi$ or $188.495\ 559$

(b) $V = \frac{1}{3}\pi r^2 h + \frac{1}{2}\frac{4}{3}\pi r^2$

$= \frac{1}{3}\pi(6^2)(8) + \frac{2}{3}\pi(6^2)$

$= \frac{288}{3}\pi + \frac{72}{3}\pi$

$= \frac{360}{3}\pi$ or $376.991\ 118$

Answers:

(a) 188 cm^2 (3sf)

(b) 377 cm^3 (3sf)

Now you practise it

A question in the style of Paper 1

A 12 cm by 12 cm ice cube tray is in the shape of a cuboid with twelve hemispheres inset as shown in the picture below. The depth of each hemisphere is 1 cm.

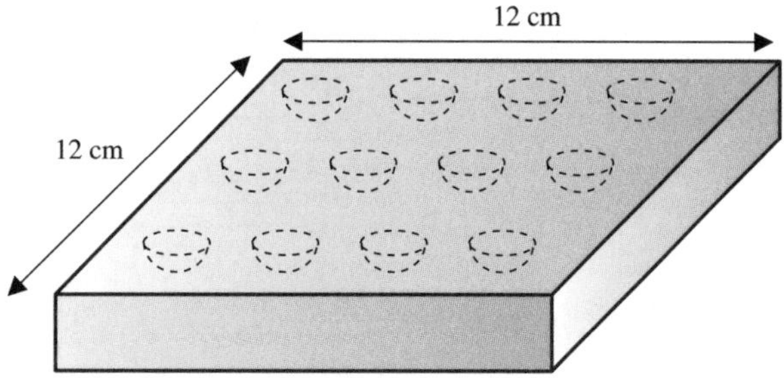

(a) If all the hemispheres were filled with water, calculate the total volume of water that the ice cube tray could hold.

It is now desired to paint the top of the ice cube tray a pleasant shade of blue.

(b) Calculate the surface area of one hemisphere on the tray.

(c) Calculate the total surface area of the top of the ice cube tray.

Working:

Answers:

(a) ....

(b) ....

(c) ....

A harder one from Specimen 2005 Paper 1

A swimming pool is in the shape of a cuboid with dimensions 50 metres long by 25 metres wide by 2.5 metres deep. If the pool is completely filled with water, calculate

(a) the volume of water in the pool;

(b) the longest straight distance a swimmer could swim in the pool;

(c) the angle the straight path in (b) makes with the end of the pool.

Working:

Answers:

(a) ....

(b) ....

(c) ....

Additional practice problems

1. From May 2007 Paper 2 *[Maximum mark: 18]*

(i) Jenny has a circular cylinder with a lid. The cylinder has height 39 **cm** and diameter 65 **mm**.

(a) Calculate the volume of the cylinder **in cm³**. Give your answer correct to **two** decimal places. *[3 marks]*

The cylinder is used for storing tennis balls.

Every ball has a **radius** of 3.25 cm.

(b) Calculate how many balls Jenny can fit in the cylinder if it is filled to the top. *[1 mark]*

(c) (i) Jenny fills the cylinder with the number of balls found in part (b) and puts the lid on. Calculate the volume of air inside the cylinder in the spaces between the tennis balls.

(ii) Convert your answer to (c) (i) into cubic metres. *[4 marks]*

(ii) An old tower (BT) leans at 10° away from the vertical (represented by line TG).

The base of the tower is at B so that angle $M\hat{B}T = 100°$.

Leonardo stands at L on flat ground 120 m away from B in the direction of the lean.

He measures the angle between the ground and the top of the tower T to be angle $B\hat{L}T = 26.5°$.

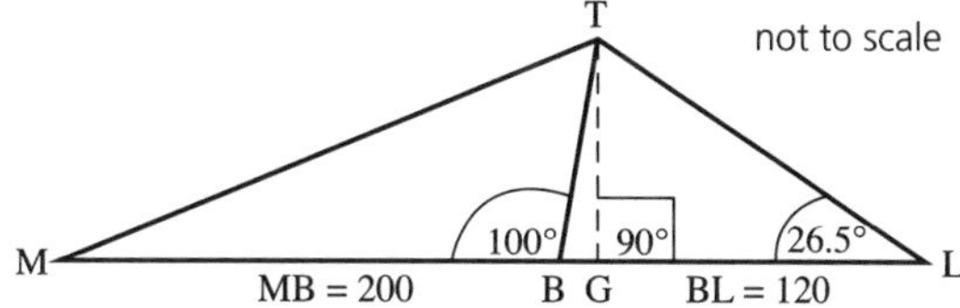

(a) (i) Find the value of angle $B\hat{T}L$.

(ii) Use the triangle BTL to calculate the sloping distance BT from the base, B to the top, T of the tower. *[5 marks]*

(b) Calculate the vertical height TG to the top of the tower. *[2 marks]*

(c) Leonardo now walks to point M, a distance 200 m from B on the opposite side of the tower. Calculate the distance from M to the top of the tower at T. *[3 marks]*

2. From November 2004 Paper 2 *[Maximum mark: 16]*

A cross-country running course is given in the diagram below. Runners start and finish at point O.

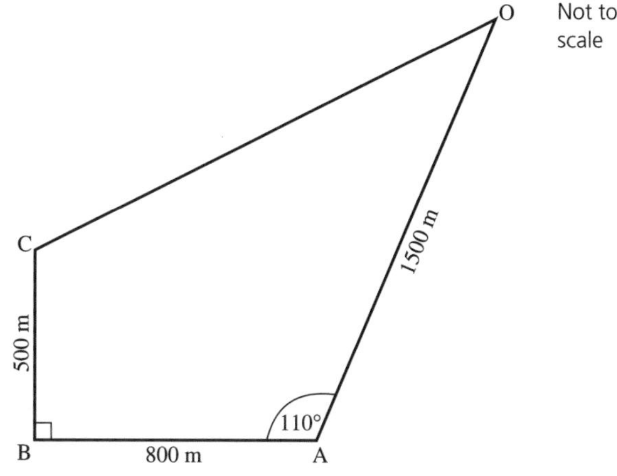

(a) Show that the distance CA is 943 m correct to 3 s.f. *[2 marks]*

(b) Show that angle BCA is 58.0° correct to 3 s.f. *[2 marks]*

(c) (i) Calculate the angle CAO.

(ii) Calculate the distance CO. *[5 marks]*

(d) Calculate the area enclosed by the course OABC. *[4 marks]*

(e) Gonzales runs at a speed of 4 ms^{-1}. Calculate the time, in minutes, taken for him to complete the course. *[3 marks]*

3. From May 2006 Paper 2 *[Maximum mark: 11]*

The figure below shows a rectangular prism with some side lengths and diagonal lengths marked. AC = 10 cm, CH = 10 cm, EH = 8 cm, AE = 8 cm.

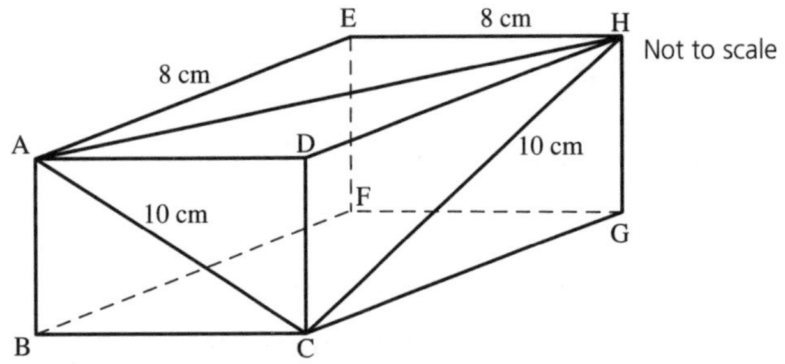

(a) Calculate the length of AH. *[2 marks]*

(b) Find the size of angle ACH. *[3 marks]*

(c) Show that the total surface area of the prism is 320 cm^2. *[3 marks]*

(d) A triangular prism is enclosed within the planes ABCD, CGHD and ABGH. Calculate the volume of this prism. *[3 marks]*

TOPIC 25

Classification of data, frequency tables and polygons

What do you need to know?

Statistical data are either *discrete* or *continuous*.

- Discrete data are those that cannot be broken up into infinitely small pieces – you have to count them as *discrete* things.
- Continuous data are those that can be broken down into infinitely many pieces.

A frequency table is a chart that simply shows how many there are of each datum (piece of information).

Frequency polygons are graphs of these data. You set up the x-axis with the variable (length of a banana, for example) and the y-axis with the frequency (how many bananas of each length).

How do you do it?

To distinguish between discrete and continuous data, you just have to look at the problem and decide whether or not the data can be written in pieces. If it cannot, like number of cars or number of desks (you can't have a fraction of a car or a fraction of a desk), it's discrete. If it can, like the amount of time to run 100 metres or the length of a car in centimetres (you can have infinitely small fractions of time or infinitely small fractions of length), it's continuous.

Frequency tables are usually set up something like this:

Length of banana	11 cm	12 cm	13 cm	14 cm
Frequency	6	8	3	9

So there are 6 bananas which are 11 cm, 8 that are 12 cm, etc.

From these data, a frequency polygon can be drawn. Plot the points from the data and then create a line graph by connecting the points. If the IB asks you to *graph* a frequency polygon, they will specify the scale and axes. If the IB asks you to *sketch* a frequency polygon, it's up to you to make a reasonable scale and axis labels.

Don't forget to label your axes and scale!

Example

On a biology field trip, you decide to measure the mass of the tusks of male warthogs (without harming the warthogs, of course). Despite the immediate danger of the task, you succeed at collecting the following data:

Mass m of tusks in kg	Frequency
13	4
13.5	7
14	3
16	11
18	5

Here is the frequency polygon:

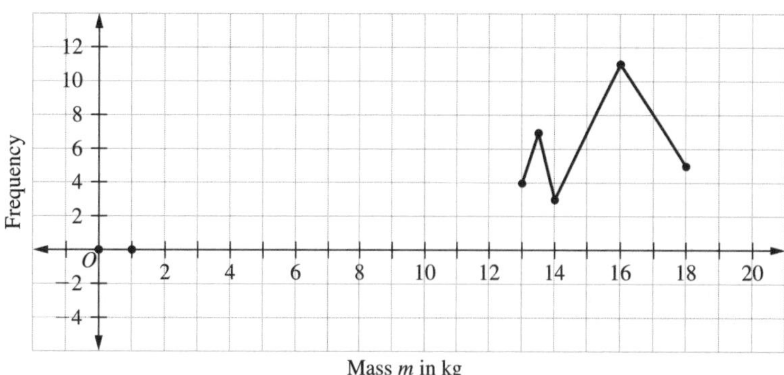

How can my calculator help me do this?

Your calculator can be helpful in graphing frequency polygons.

Entering the data in L1 and L2 in your statistics editor for the example above:

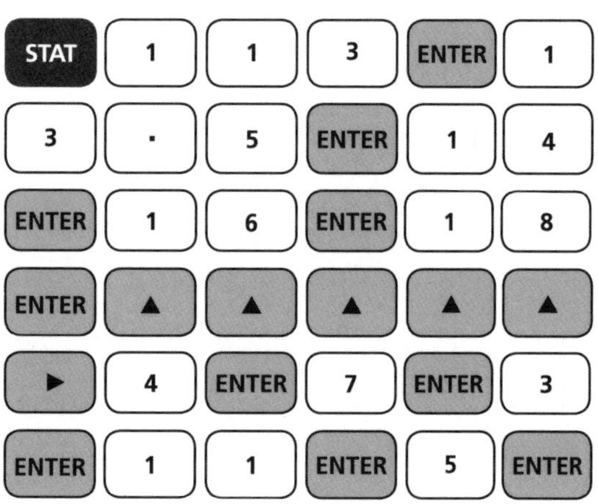

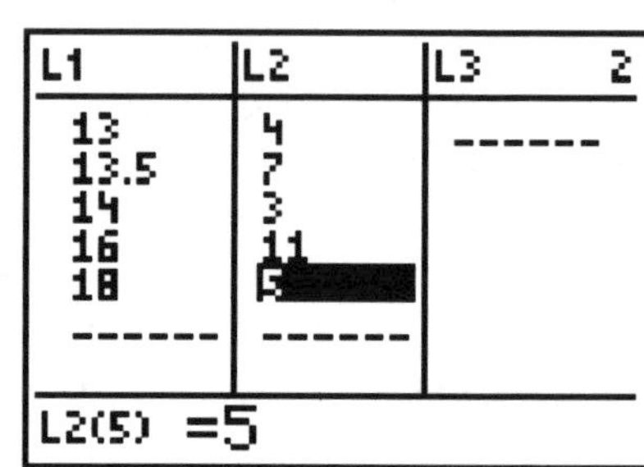

Creating the graph for the example above:

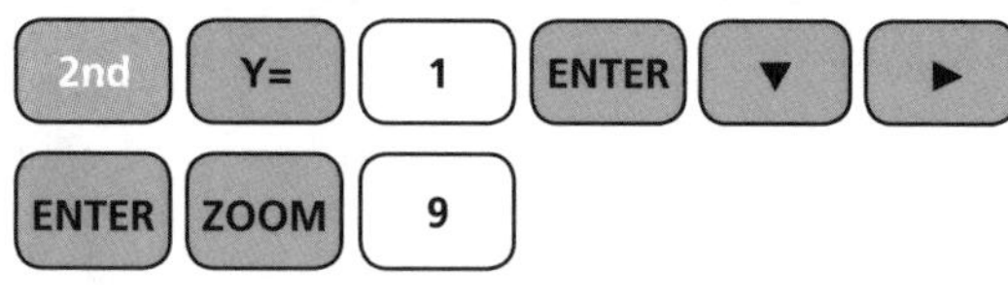

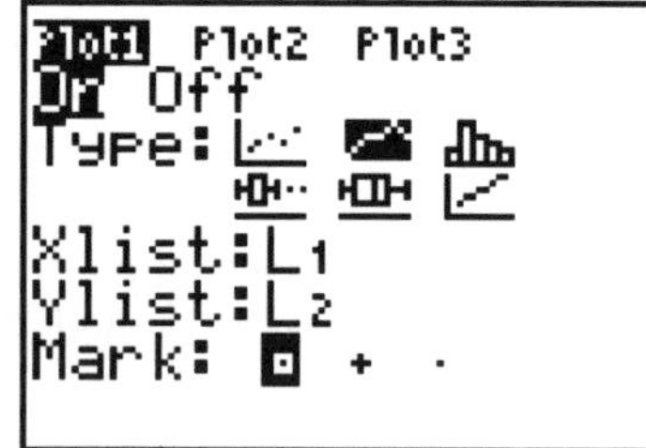

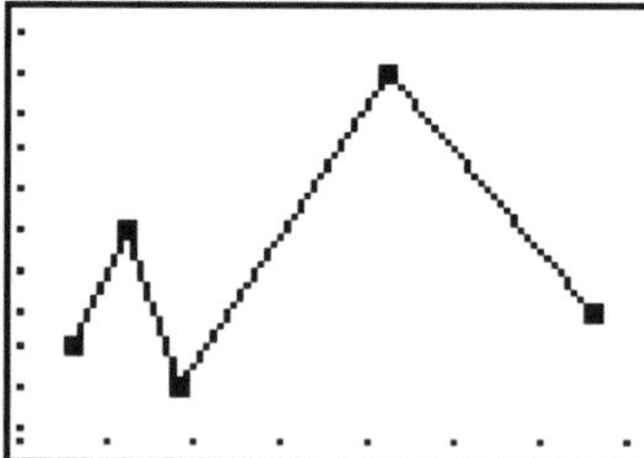

Examples from past IB papers

An easy one from May 2007 Paper 1

(you might need to read the next couple of sections first)

(a) State which of the following sets of data are discrete.

(i) Speeds of cars travelling along a road

(ii) Numbers of members in families

(iii) Maximum daily temperatures

(iv) Heights of people in a class measured to the nearest cm

(v) Daily intake of protein by members of a sporting team.

The boxplot below shows the statistics for a set of data.

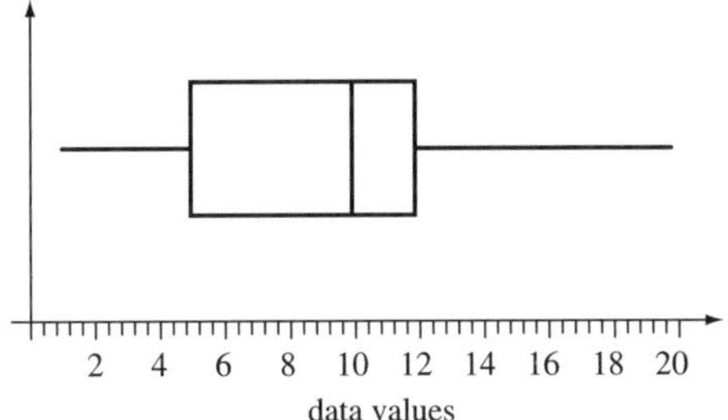

(b) For this data set write down the value of

(i) the median

(ii) the upper quartile

(iii) the minimum value present.

(c) Write down three different integers whose mean is 10.

Working:

(c) $\frac{(8 + 10 + 12)}{3} = \frac{30}{3} = 10$

(lots of possible answers here!)

Answers:

(a) (ii) and (iv) are discrete

(b) (i) 10

(ii) 12

(iii) 1

(c) 8, 10, 12

Now you practise it

From November 2005 Paper 1 (modified)

The diagram shows the frequency histogram for the number of petals on roses.

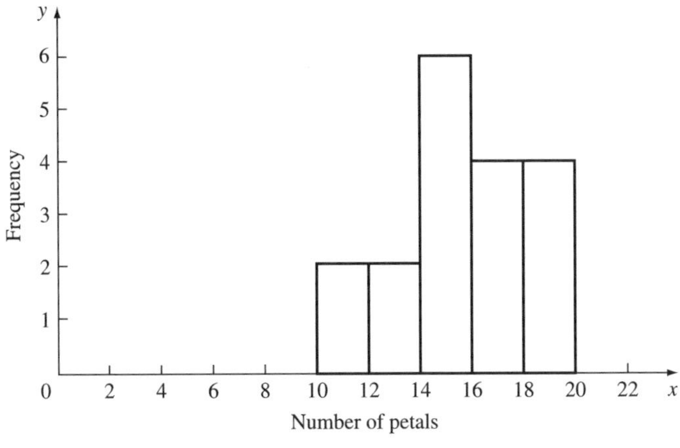

The information is displayed in the frequency table below.

Number of petals (x)	Frequency
$10 < x \leq 12$	2
$12 < x \leq 14$	a
$14 < x \leq 16$	6
$16 < x \leq 18$	b
$18 < x \leq 20$	4

(a) Find the values of a and b.

(b) State the modal group.

(c) Calculate an estimate of the mean number of petals.

Working:

Answers:

(a) ....

(b) ....

(c) ....

TOPIC 26

Grouped data, histograms, stem-and-leaf plots

What do you need to know?

Grouped data is used when you have too much data to present on one graph all at once. It is a technique that shows the data in terms of categories, rather than individually. A graph of grouped data that shows how many are in each equally-sized category is called a *histogram*.

A table that shows all the data with the last digits separated from the rest is called a *stem-and-leaf plot*.

How do you do it?

To group data, you place the data into different categories, e.g. $4 \leq x < 7$ or $3.1 < x < 3.2$. The categories do not have to be the same size but usually they are. Then you tally up the number of data in each category. This total is called its *frequency*.

IRENE SAYS:

Warning! Don't write intervals like this: 3–5, 6–8, ... This is not precise enough and often leaves gaps. (Where would you put 5.5?) Write your intervals the proper way or you may lose marks on the exam!

To create a histogram, you draw a bar chart with the intervals on the *x*-axis and the frequency on the *y*-axis. You do not need to start the *x*-axis at zero but you must clearly label the axes and indicate scale.

SCOTT SAYS:

The bars of a histogram must touch one another. If there is a gap in the intervals, you must make the bars meet. For example, for the intervals $3 \leq x < 5$ and $6 \leq x < 8$, you should make the first bar from 2.5 to 5.5 and the second from 5.5 to 8.5.

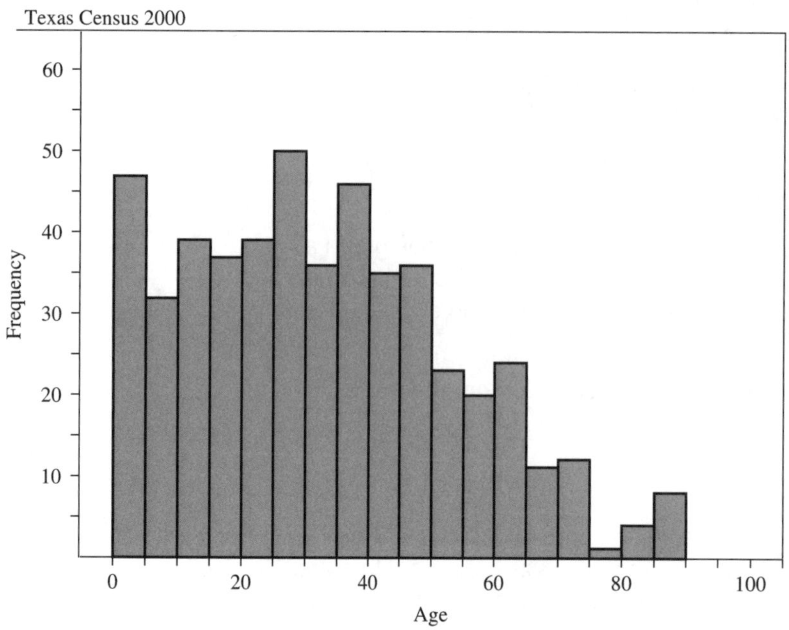

To create a stem-and-leaf plot, separate the last digit of each data item from the rest of the number. The last digit becomes the *leaf* and the rest becomes the *stem*.

123 becomes 12 | 3   0.0923 becomes 0.092 | 3

Example

Data are collected from 36 IB students concerning the number of hours needed to write an IB Extended Essay rough draft. The data are below.

12	34	19	27	41	22	18	36	24	21	11	32
28	31	26	40	20	16	22	29	33	12	25	22
18	54	30	22	21	18	28	27	34	48	20	12

Here is the stem-and-leaf plot.

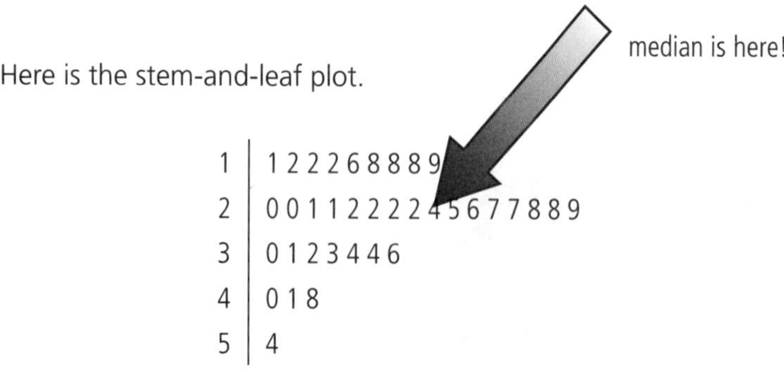

Note that this organised picture, besides giving a good idea of the distribution of the data, allows you to find the median. In this example, the median is between the 18th and 19th numbers (there are 36 numbers), so you can count from the top or bottom and find them – they are 24 and 25. So the median is 24.5.

Here is one way the data could be grouped.

Hours	Frequency
$10 \le x < 15$	4
$15 \le x < 20$	5
$20 \le x < 25$	9
$25 \le x < 30$	7
$30 \le x < 35$	6
$35 \le x < 40$	1
$40 \le x < 45$	2
$45 \le x < 50$	1
$50 \le x < 55$	1

And here's the resulting histogram.

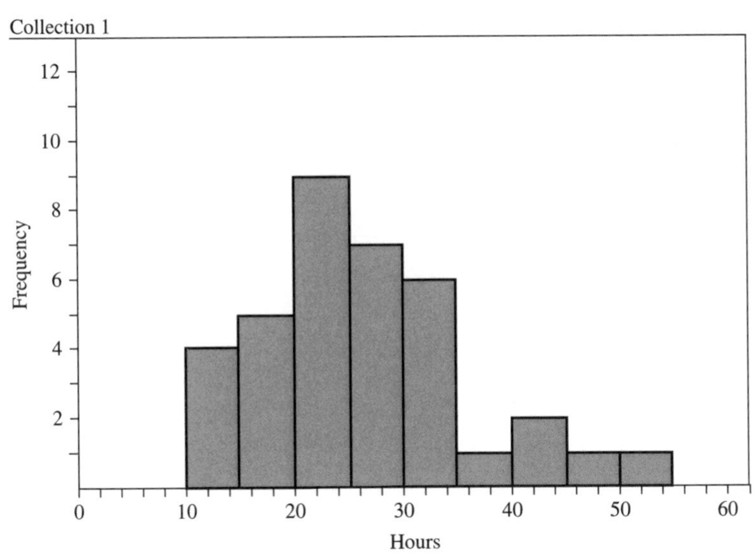

How can my calculator help me do this?

Your calculator does not do stem-and-leaf plots but does sort your data nicely, and it does histograms very well.

Entering the data from the example above in L1 and sorting the data for a stem-and-leaf plot:

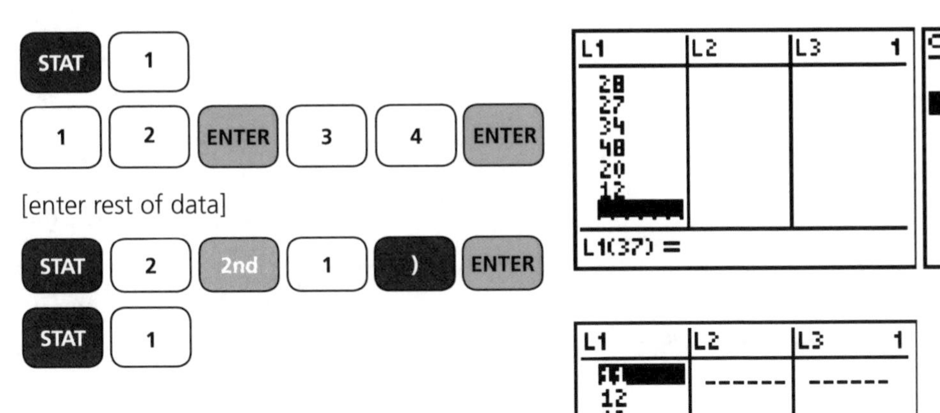

Creating the histogram for the example above:

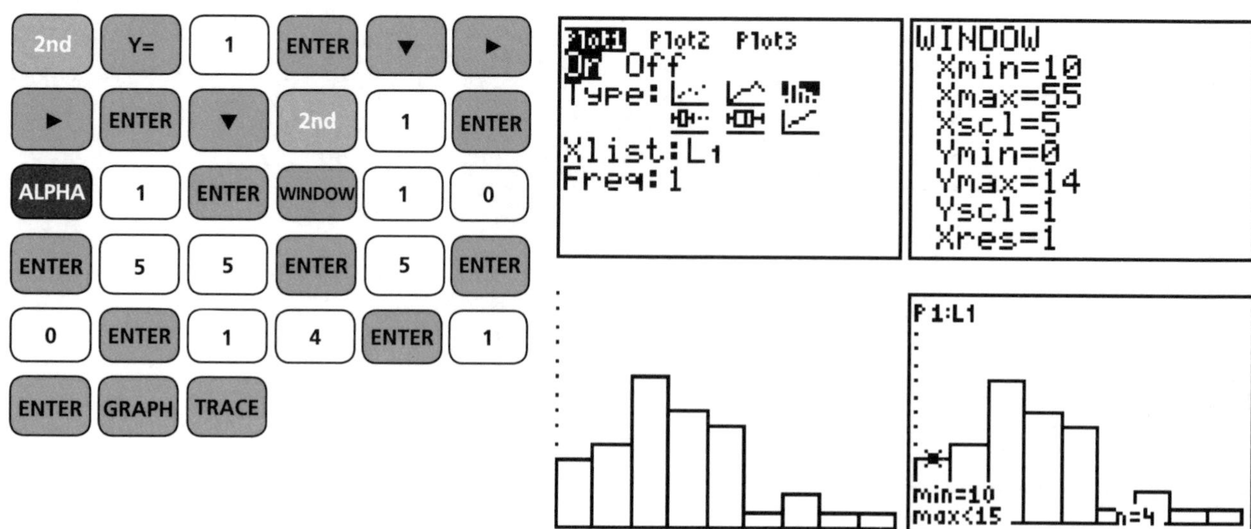

Examples from past IB papers

An easy one from November 2007 Paper 1

The birth weights, in kilograms, of 27 babies are given in the diagram below.

		key 1|7 = 1.7 kg
1	7, 8, 9	
2	1, 2, 2, 3, 5, 5, 7, 8, 9	
3	0, 1, 3, 4, 5, 5, 6, 6, 7, 9	
4	1, 1, 2, 3, 7	

(a) Calculate the mean birth weight.

(b) Write down:

(i) the median weight,

(ii) the upper quartile.

Working:

(a) $\frac{(1.7 + 1.8 + 1.9 + 2.1 + 2.2 + \ldots + 4.7)}{27} = 3.096\ 296$

Answers:

(a) 3.10 kg (3sf)

(b) (i) 3.1 kg

(ii) 3.7 kg

Now you practise it

An easy one from November 2003 Paper 1 (modified)

The frequency histogram below shows the times, in minutes, taken by students to complete an assignment.

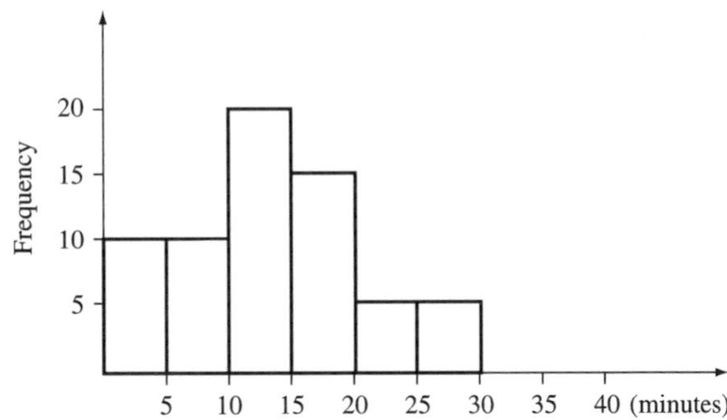

The histogram shows that 20 students took between 10 and 15 minutes to complete the assignment.

(a) How many students took between 15 and 20 minutes?

(b) How many students took between 15 and 30 minutes?

(c) 15 students took between 30 and 35 minutes. Complete the histogram above.

Working:

Answers:

(a) ....

(b) ....

Cumulative frequencies, box-and-whisker plots

What do you need to know?

Cumulative frequencies are sums of the frequencies of groups of data. You use them to create a *cumulative frequency curve* (also called an *ogive*) and from this you can find the median, quartiles, interquartile range (IQR) and percentiles of the data.

A box-and-whisker plot is another way to graph data in order to show the median, quartiles, IQR and the max and min of the data.

For definitions of median, quartiles, IQR and percentiles, see the next section.

How do you do it?

In order to make a cumulative frequency graph, you need to make a table first. The table looks very similar to the one you make for a histogram, but you add another column to it called "cumulative frequency". Start with the first group and rewrite its frequency in the "cumulative" column. In the next group, add its frequency to the previous frequency and write this sum in the "cumulative" column and so on.

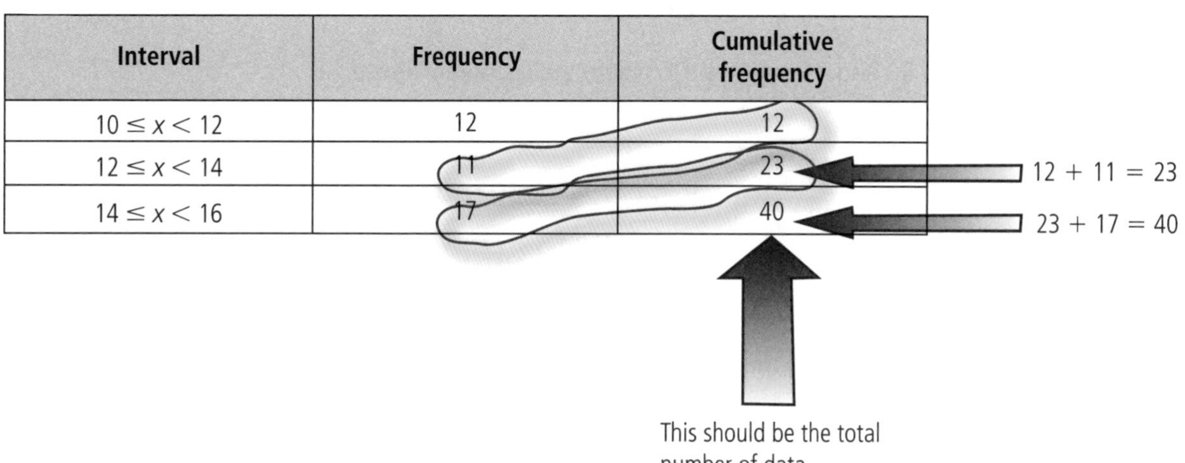

Interval	Frequency	Cumulative frequency
$10 \le x < 12$	12	12
$12 \le x < 14$	11	23
$14 \le x < 16$	17	40

Now draw a cumulative frequency histogram.

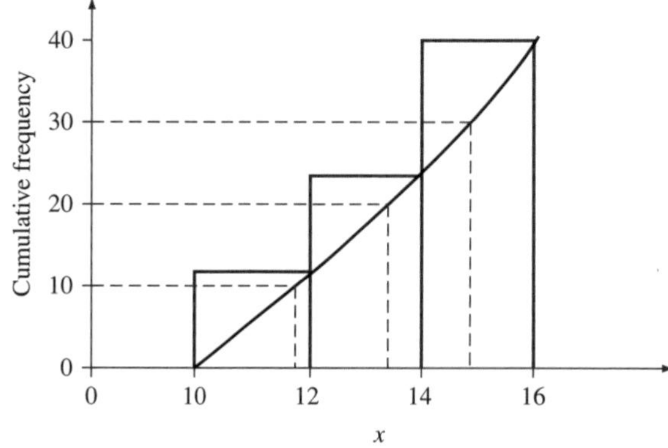

When you draw your curve, start with the *lower left-hand corner of the first bar* and then connect the *upper right-hand corners* of each bar. You don't actually have to draw the bars but it usually helps.

SCOTT SAYS:

Make sure your curve does not go through (0, 0) unless it is part of the data! This is a very common mistake.

To find the median, go halfway up the y-axis, over to the curve, and down to the x-axis. To find the 1st and 3rd quartiles, go up $\frac{1}{4}$ and $\frac{3}{4}$ up the y-axis, over to the curve, and again back down to the x-axis.

To make a box-and-whisker plot, draw a box so that the sides of the box align with the 1st and 3rd quartiles. Draw a vertical line in the box that aligns with the median. Draw "whiskers" out to the max and min from the box.

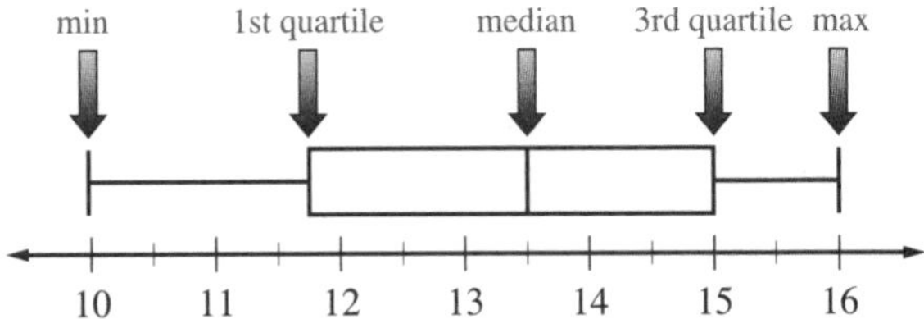

Example

Data are gathered about the number of eggs laid by the endangered giant turtle *Rafetus swinhoei* or the Shanghai soft-shell turtle.

Number of eggs per turtle	Frequency (number of turtles)	Cumulative frequency
$40 \leq x < 50$	7	7
$50 \leq x < 60$	9	16
$60 \leq x < 70$	14	30
$70 \leq x < 80$	12	42
$80 \leq x < 90$	10	52
$90 \leq x < 100$	8	60

A cumulative frequency curve is created to find the median, 1st quartile and 3rd quartile.

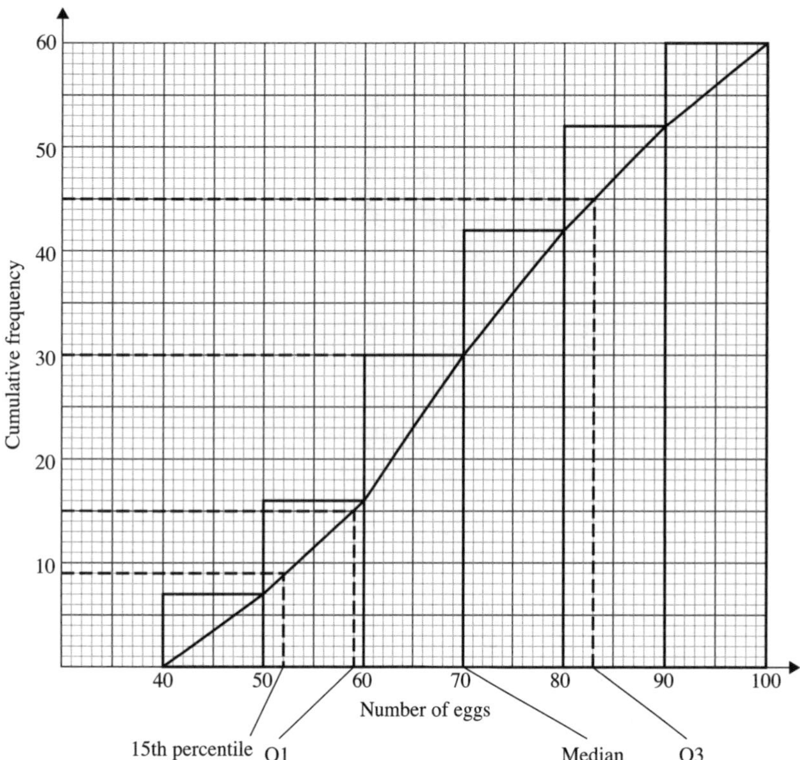

So the median is approximately 70 eggs, the 1st quartile is approximately 59 eggs and the 3rd quartile is approximately 83 eggs. Therefore the IQR = 83 − 59 = 24 eggs.

If you want to find the 15th percentile, you calculate 15% of the total number of turtles surveyed and find that point on the curve:

15% of 60 = $0.15 \times 60 = 9$. 9 on the *y*-axis corresponds to 52 on the *x*-axis so the 15th percentile is approximately 52 eggs.

Now suppose we know that the smallest number of eggs was 41 and the greatest number of eggs was 98. A box-and-whisker plot can now be drawn.

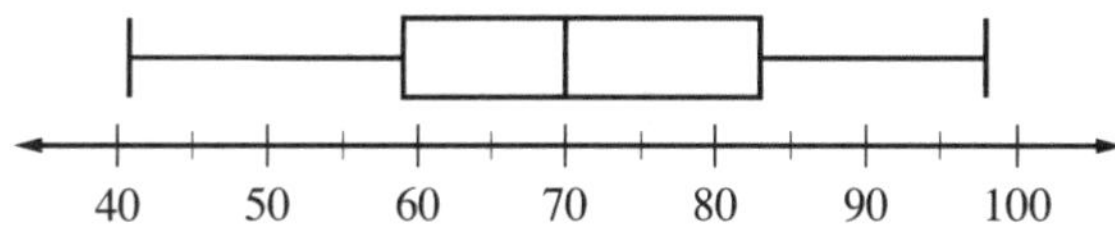

How can my calculator help me do this?

Your calculator is not useful for cumulative frequency curves (it cannot draw the curve) but it can draw box-and-whisker plots *if you have the original data*. If you have an example like the one above, the GDC is not helpful.

Drawing a box-and-whisker plot given data 3, 7, 9, 10, 10, 12, 14, 16, 19, 19, 19, 22:

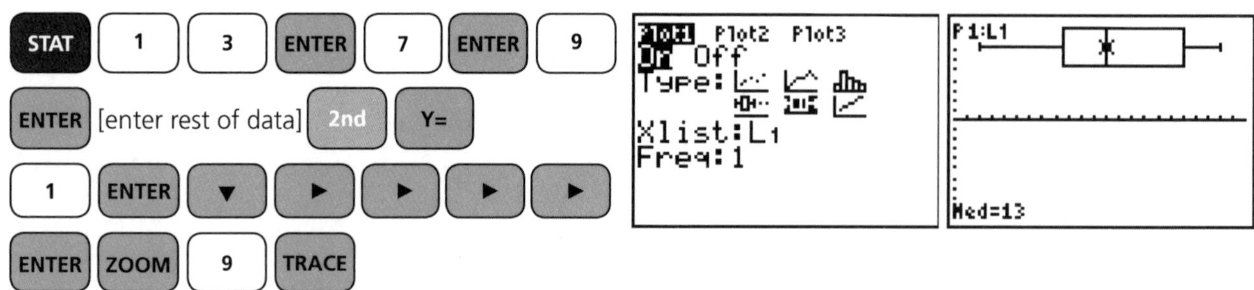

Examples from past IB papers

An easy one from May 2006 Paper 1

The heights (cm) of seedlings in a sample are shown below.

6	3,	7			
7	2,	5,	8		
8	3,	6,	6,	8,	8
9	2,	5,	7,	8	
10	3,	6,	6		
11	2,	2			

key 6 | 3 represents 63 cm

(a) State how many seedlings are in the sample.

(b) Write down the values of

(i) the median,

(ii) the first and third quartiles.

(c) Calculate the range.

(d) Using the scale below, draw a box and whisker plot for this data.

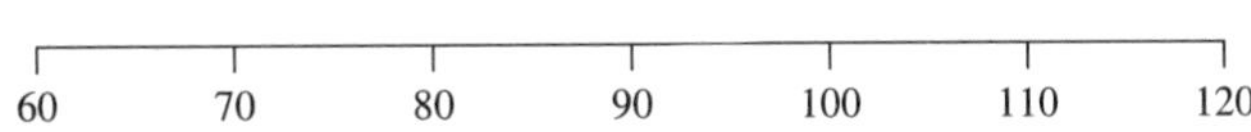

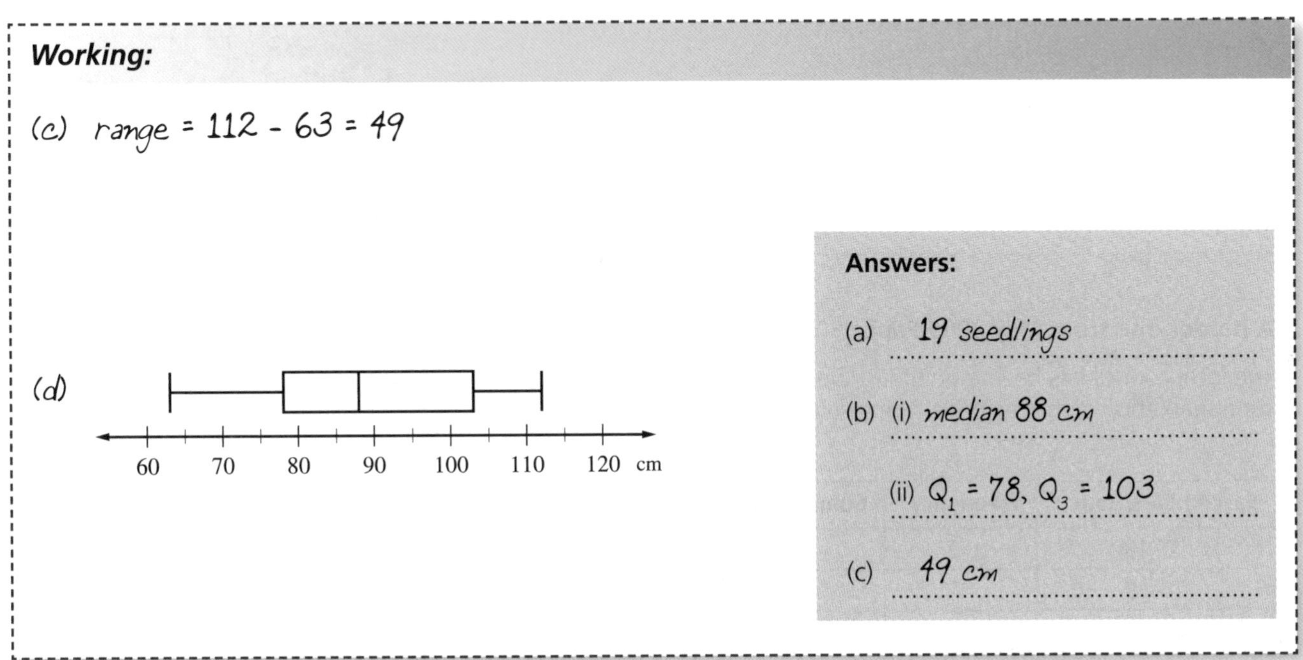

Now you practise it

An easy one from May 2004 Paper 1

The cumulative frequency table below shows the ages of 200 students at a college.

Age	Number of Students	Cumulative Frequency
17	3	3
18	72	75
19	62	137
20	31	*m*
21	12	180
22	9	189
23–25	5	194
>25	6	*n*

(a) What are the values of *m* and *n*?

(b) How many students are younger than 20?

(c) Find the value in years of the lower quartile.

Working:

Answers:

(a) ....

(b) ....

(c) ....

A harder one from May 2005 Paper 1

The local council has been monitoring the number of cars parked near a supermarket on an hourly basis. The results are displayed below.

Parked Cars/Hour	Frequency	Cumulative Frequency
0–19	3	3
20–39	15	18
40–59	25	*w*
60–79	35	78
80–99	17	95

(a) Write down the value of *w*.

(b) Draw and label the **cumulative frequency** graph for this data.

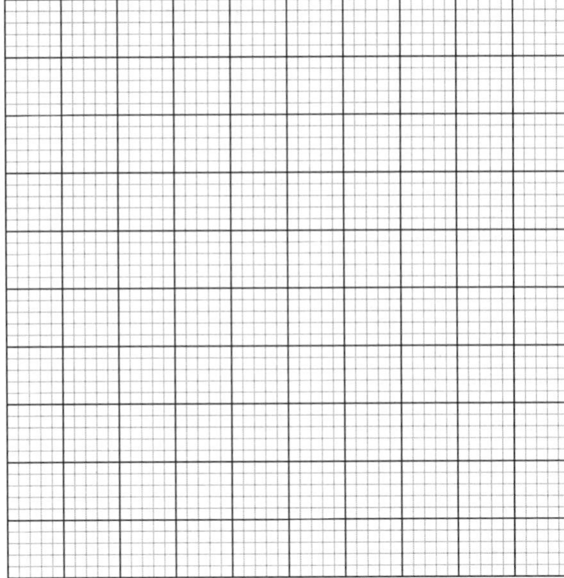

(c) Determine the median number of cars per hour parked near the supermarket.

Working:

Answers:

(a) ....

(c) ....

TOPIC 28

1-variable statistical calculations

What do you need to know?

Several calculations can provide insight into a set of data. 1-variable calculations are used when you have only one kind of data, e.g. length of a hand in cm, time it takes to eat a banana, etc. You are not comparing one data set to another – this comes in the next section. The calculations are mean, median, mode, modal group, range, quartiles, interquartile range (IQR) and standard deviation. The definitions are:

mean ($\bar{x}$)	average (add up and divide by number of data)
approx mean of grouped data	the average when you have grouped data
median	middle value
mode	the most frequently occurring value
modal group	when you have grouped data, the group that occurs most frequently
range	the difference between the highest value and the lowest value
1st quartile (q_1)	the median of the values below the median of the whole data set
2nd quartile	same as median of the whole data set – also known as the 50th percentile
3rd quartile (q_3)	the median of the values above the median of the whole data set
interquartile range (IQR)	the difference between the 3rd quartile and the 1st quartile
standard deviation (s_x)	a measure of how spread out the data are – approximately 68% of the data are within one standard deviation of the mean

How do you do it?

mean	add up the data values and divide by the number of values
approximate mean of grouped data	found by taking the middle value of each group, multiplying by the number of values in each group and dividing by the total number of values
median	line up data values in order and find the middle value – if there are an even number of values, you find the average of the two middle values
mode	find the value that occurs the most often
modal group	choose the group with the highest frequency
range	subtract the lowest value from the highest value
1st quartile (q_1)	line up data values in order from smallest to largest and find the middle value of all the numbers to the left of the median – if there are an even number of values, you find the average of the two middle values
3rd quartile (q_3)	line up data values in order from smallest to largest and find the middle value of all the numbers to the right of the median – if there are an even number of values, you find the average of the two middle values
interquartile range (IQR)	subtract the 1st quartile from the 3rd quartile
standard deviation (s_x)	use the formula in your booklet or use your GDC (see below)

Example

The continuous data set below shows the amount of time that 10 people took to eat a hamburger in seconds:

45.2 33.1 71.2 75.4 80.0 55.9 73.1 50.8 49.6 81.2

Mean is $\bar{x} = \dfrac{45.2 + 33.1 + 71.2 + 75.4 + 80.0 + 55.9 + 73.1 + 50.8 + 49.6 + 81.2}{10}$

$= 61.55$ sec $= 61.6$ sec (3sf)

To find the median and other values, first sort the data values from smallest to largest:

33.1 45.2 49.6 50.8 55.9 71.2 73.1 75.4 80.0 81.2

The median is the average of 55.9 and 71.2 $= \frac{55.9 + 71.2}{2} = 63.55$ sec $= 63.6$ sec (3sf)

There is no mode as all values are different (but there is a modal group – see below).

1st quartile is the median of the lower five values which is 49.6 sec.

3rd quartile is the median of the upper five values which is 75.4 sec.

IQR is $75.4 - 49.6 = 25.8$ sec

Standard deviation is found via the formula in the booklet:

6.6	Standard deviation	$s_n = \sqrt{\frac{\sum_{i=1}^{k} f_i (x_i - \bar{x})^2}{n}}$, where $n = \sum_{i=1}^{k} f_i$

So $s_x = \sqrt{\frac{(45.2 - 61.55)^2 + (33.1 - 61.55)^2 + (71.2 - 61.55)^2 + \ldots + (81.2 - 61.55)^2}{10}}$

$= 15.848 \ldots$ sec $= 15.9$ sec (3sf)

Therefore 68% of the data should lie between 15.8 seconds above the mean and 15.8 seconds below the mean, or between 45.8 seconds and 77.4 seconds.

If the data are presented in groups, you can find the grouped mean and modal group:

$30 \leq x < 40$	$40 \leq x < 50$	$50 \leq x < 60$	$60 \leq x < 70$	$70 \leq x < 80$	$80 \leq x < 90$
1	2	2	0	3	2

Grouped mean is $\bar{x} = \frac{35 \cdot 1 + 45 \cdot 2 + 55 \cdot 2 + 65 \cdot 0 + 75 \cdot 3 + 85 \cdot 2}{10}$

$= 63.0$ sec

The modal group is $70 \leq x < 80$ as this group has the highest frequency.

How can my calculator help me do this?

Your GDC does all 1-variable calculations for you in one shot! If your Stat Editor is a mess from another problem, reset the editor before moving on (see Hints section at the beginning of this book).

Finding the 1-variable statistics for the data in the example above:

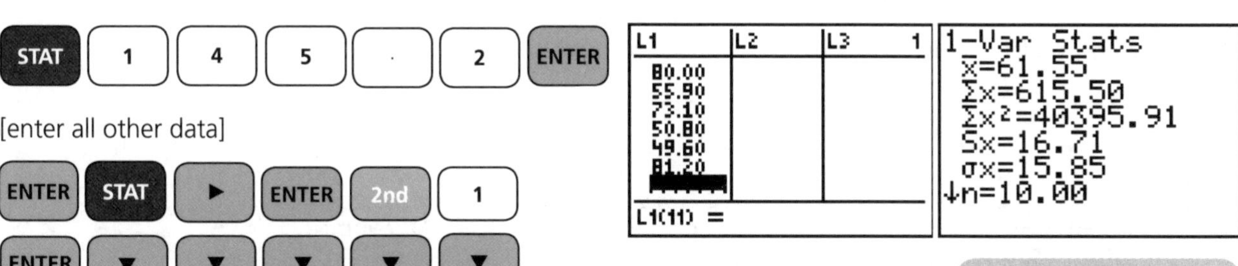

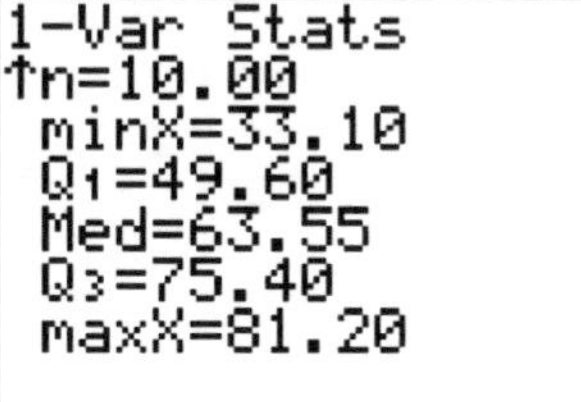

WARNING!
The correct standard deviation to use on the GDC is the σx value, not the Sx value. Make sure you use the correct one!

If you want to use grouped data with frequencies (like in the bottom part of the example), you use both lists L1 and L2 in the GDC:

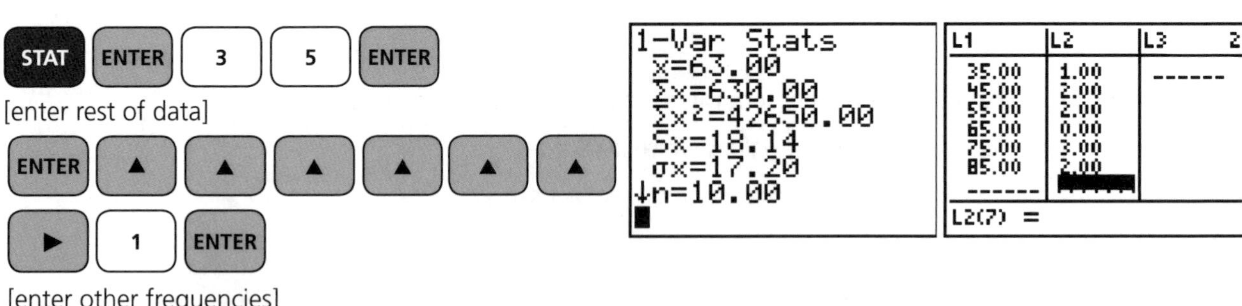

Examples from past IB papers

An easy one from May 2005 Paper 1

The numbers of games played in each set of a tennis tournament were

9, 7, 8, 11, 9, 6, 10, 8, 12, 6, 8, 13, 7, 9, 10, 9, 10, 11, 12, 8, 7, 13, 10, 7, 7

The raw data has been organised in the frequency table below.

games	frequency
6	2
7	5
8	n
9	4
10	4
11	2
12	2
13	2

(a) Write down the value of n.

(b) Calculate the mean number of games played per set.

(c) What percentage of the sets had more than 10 games?

(d) What is the modal number of games?

Working:

(b) $$\frac{(6(2) + 7(5) + 8(4) + 9(4) + 10(4) + 11(2) + 12(2) + 13(2))}{(2 + 5 + 4 + 4 + 4 + 2 + 2 + 2)}$$

$$= \frac{227}{25} = 9.08$$

(c) $\frac{6}{25} \times 100 = 24\%$

Answers:

(a) $n = 4$

(b) mean = 9.08 games

(c) 24%

(d) mode = 7 games

A harder one, also from May 2005 Paper 1

Peter has marked 80 exam scripts. He has calculated the mean mark for the scripts to be 62.1. Maria has marked 60 scripts with a mean mark of 56.8.

(a) Peter discovers an error in his marking. He gives two extra marks each to eleven of the scripts. Calculate the new value of the mean for Peter's scripts.

(b) After the corrections have been made and the marks changed, Peter and Maria put all their scripts together. Calculate the value of the mean for all the scripts.

Working:

(a) $80 \times 62.1 + 11 \times 2 = 4990$

$\frac{4990}{80} = 62.375$

(b) $4990 + 60 \times 56.8 = 8398$

$\frac{8398}{140} = 59.985\,714$

Answers:

(a) 62.4 (3sf)

(b) 60.0 (3sf)

Now you practise it

An easy one from November 2004 Paper 1

The number of hours that a professional footballer trains each day in the month of June is represented in the following histogram.

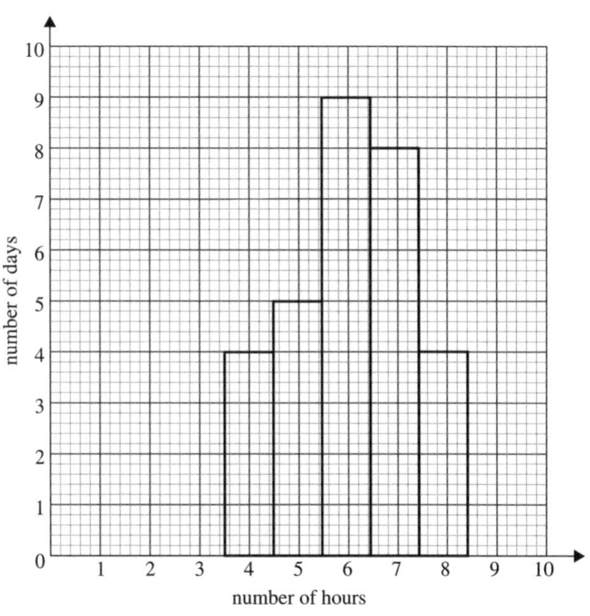

(a) Write down the modal number of hours trained each day.

(b) Calculate the mean number of hours he trains each day.

Working:

Answers:

(a) ....

(b) ....

A harder one from May 2004 Paper 1

A group of students has measured the heights of 90 trees. The class calculate the mean height to be $\bar{x} = 12.4$ m with standard deviation $s = 5.35$ m. One student notices that two of the measurements, 44.5 m and 43.2 m, are much too big and must be wrong.

(a) How many standard deviations away from the mean of 12.4 is the value 44.5?

The incorrect measurements of 44.5 m and 43.2 m must be removed from the data.

(b) Calculate the new value of $\bar{x}$ after removing the two unwanted values.

Working:

Answers:

(a) ....

(b) ....

TOPIC 29 Linear regression (line of best fit)

What do you need to know?

A regression line (line of best fit) is a straight line that attempts to show a trend with a set of 2-variable data. This trend could be a positive trend (e.g. the more hours you spend studying, the greater your maths test grade – the line has a positive slope) or a negative trend (e.g. the more hours you spend studying, the smaller number of hours of sleep you get – the line has a negative slope).

The closer the data are to a line of best fit, the greater the *Pearson product-moment correlation coefficient* (otherwise known as "r").

- If the data are positive and a perfect fit, $r = 1$.
- If the data are negative and a perfect fit, $r = -1$.
- If there is absolutely no relationship between the data at all, $r = 0$.
- A strong fit is one where r is above 0.9 or below -0.9.
- r cannot be greater than 1 or less than -1!

All lines of best fit go through the coordinates of the means. In other words, the line of best fit always goes through the point $(\bar{x}, \bar{y})$, where $\bar{x}$ is the mean of the x-axis data, and $\bar{y}$ is the mean of the y-axis data.

The equation of the line of best fit is an equation of a straight line $y = mx + c$. You use this equation to make *predictions* about other points near the existing data.

You cannot use the equation to make predictions far away from the existing data!

How do you do it?

To find the equation of the line of best fit or r, you can either use the formulae in your booklet or use your GDC.

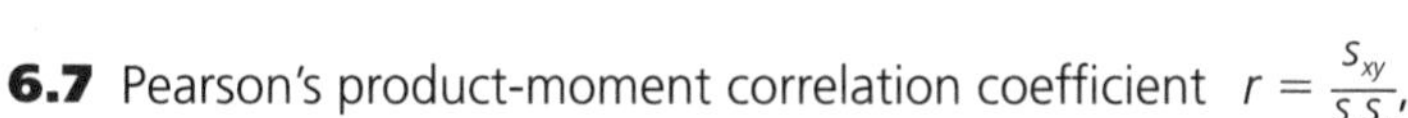

6.7 Pearson's product-moment correlation coefficient $r = \frac{s_{xy}}{s_x s_y}$,

where $s_x = \sqrt{\frac{\sum_{i=1}^{n}(x_i - \bar{x})^2}{n}}$, $s_y = \sqrt{\frac{\sum_{i=1}^{n}(y_i - \bar{y})^2}{n}}$

and s_{xy} is the covariance

6.8 Equation of regression line for y on x $\quad y - \bar{y} = \frac{s_{xy}}{s_x^2}(x - \bar{x})$

To make predictions, substitute a number for either x or y, depending on what you have, and solve the equation for what you need.

To graph a line of best fit, plot the data points on a graph first (accurately!), plot the point $(\bar{x}, \bar{y})$ and draw the line of best fit through this point, using the correct slope.

Example

A group of IB students perform a study to see if there is a mathematical relationship between the number of hours of sleep IB students get per week and their ability to memorise IB biology terminology. Here are their results looking at 12 people.

Hours of sleep per week (x)	42	47	39	45	54	51	48	31	44	50	49	46
Number of biology words memorised (y)	107	119	92	115	131	134	123	76	100	125	118	112

If we are given the covariance (it has to be given in IB exams), we can use the formula in the booklet to find the equation of the line of best fit.

Covariansce $s_{xy} = 90.5$ (given), $\bar{x} = 45.5$, $\bar{y} = 112.67$, $s_x = 5.8523$ (from GDC)

So $y - \bar{y} = \frac{s_{xy}}{s_x^2}(x - \bar{x})$

$y - 112.67 = \frac{90.5}{(5.8523)^2}(x - 45.5)$

$y - 112.67 = 2.64238 \cdot (x - 45.5)$

$y - 112.67 = 2.64238x - 120.228$

$y = 2.64x - 7.56$ (3sf)

You can get this much faster from the GDC – see below.

Pearson coefficient calculation:

$s_{xy} = 90.5$, $s_x = 5.8523$, $s_y = 16.0433$ (from GDC)

$r = \frac{s_{xy}}{s_x s_y} = \frac{90.5}{5.8523 \cdot 16.0433} = 0.964$ (3sf)

This also is obtained much faster on the GDC – see below.

Now we can make a scatterplot of the data (note the labelling of the axes and the indication of scale!) and draw the line of best fit.

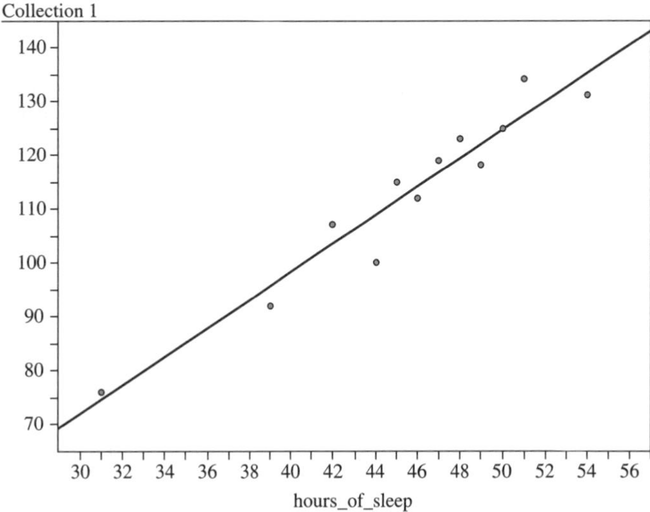

Note that this is a *strong* and *positive* trend.

How can my calculator help me do this?

This is another area where the GDC is key! Make sure you know how to do this. If your Stat Editor is a mess from another problem, reset the editor before moving on (see Hints section at the beginning of this book).

Entering the data into the GDC and getting the 1-variable statistics for both x and y:

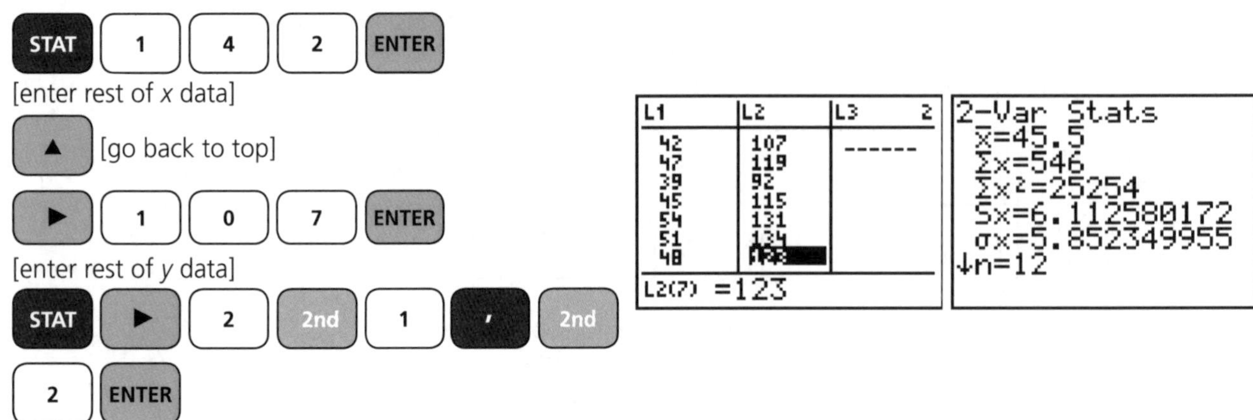

Getting the equation of the line of best fit and r value, making a scatterplot and drawing the line on the graph:

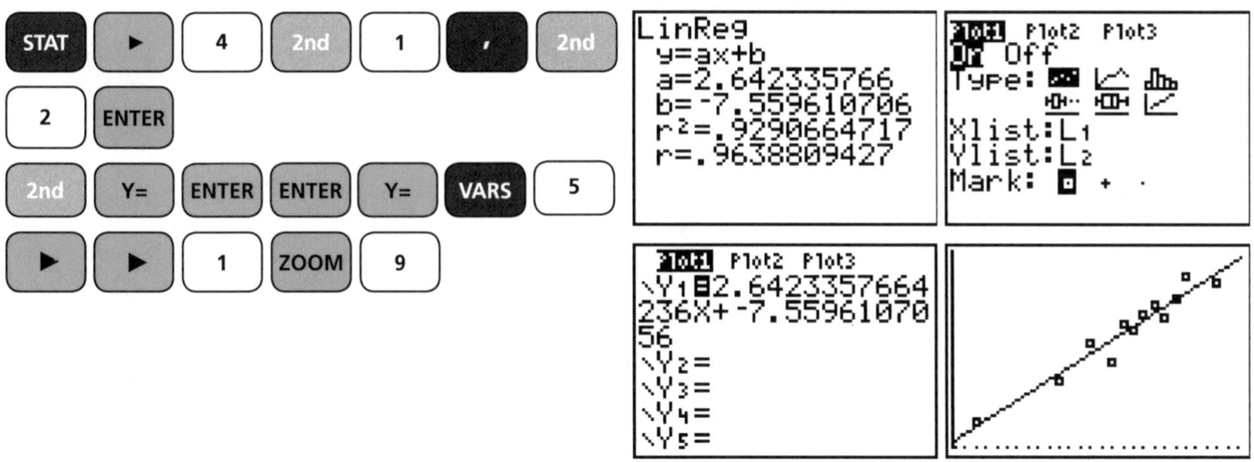

Examples from past IB papers

An easy one from May 2003 Paper 1

A group of 15 students was given a test on mathematics. The students then played a computer game. The diagram below shows the scores on the test and the game.

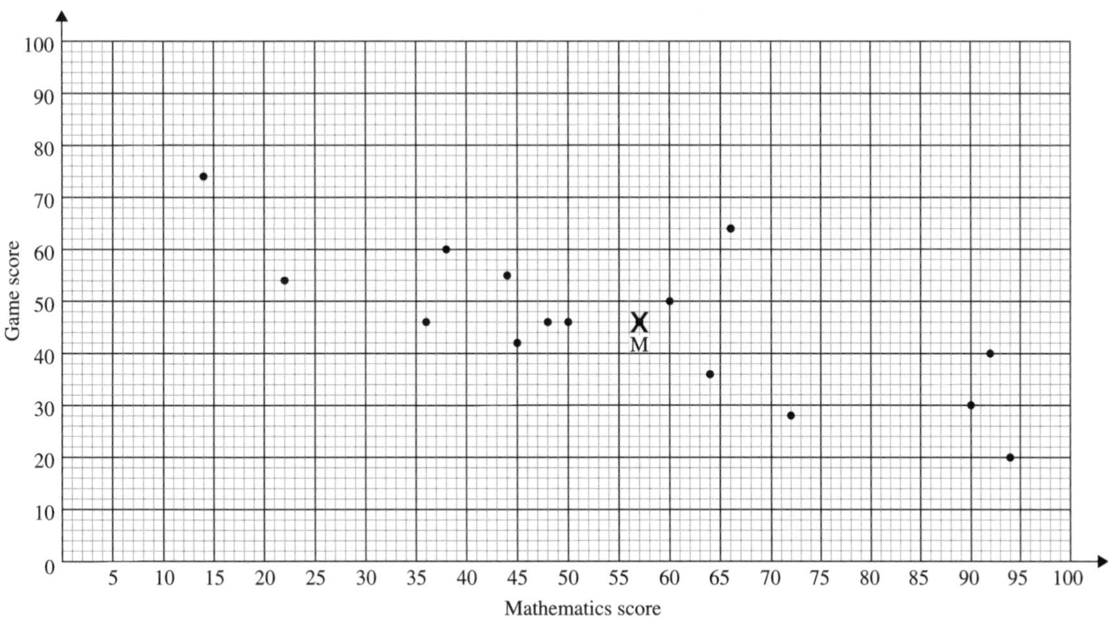

The mean score on the mathematics test was 56.9 and the mean score for the computer game was 45.9. The point M has coordinates (56.9, 45.9).

(a) Describe the relationship between the two sets of scores.

A straight line of best fit passes through the point (0, 69).

(b) On the diagram draw this straight line of best fit.

Jane took the tests late and scored 45 at mathematics.

(c) Using your graph or otherwise, estimate the score Jane expects on the computer game, giving your answer to the nearest whole number.

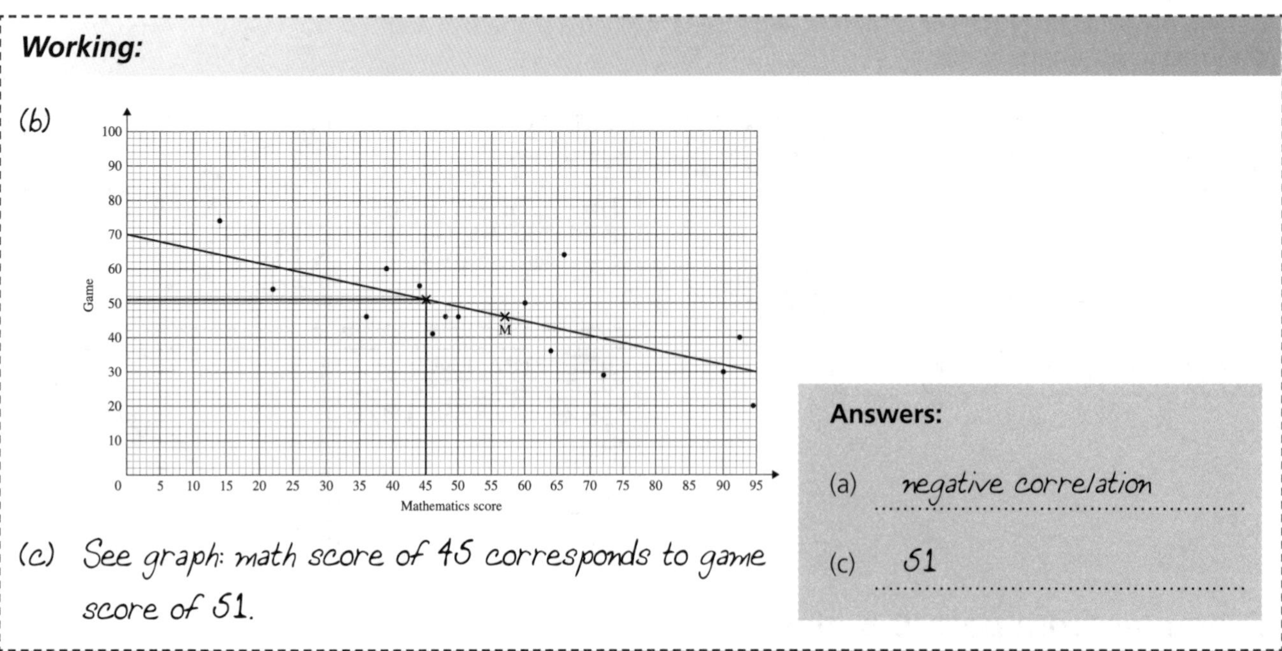

A harder one from May 2007 Paper 2 (modified)

A fisherman weighs a sample of 15 of the fish caught in his net one morning. The weight W was plotted against length L as shown below.

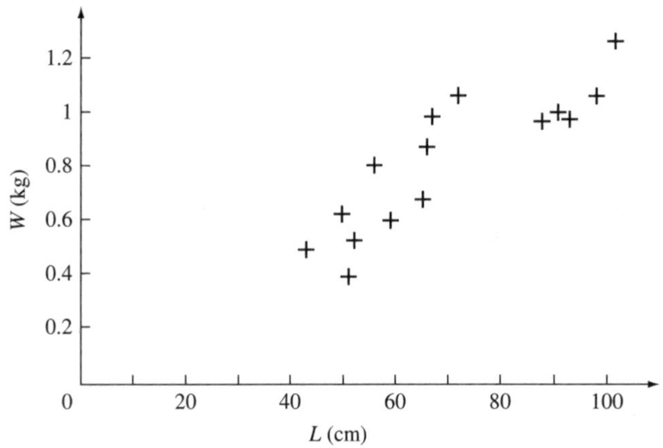

(a, b) Exactly **two** of the following statements about the plot could be correct. Identify the two correct statements.

Note: You do **not** need to enter data in a GDC **or** to calculate r exactly.

(i) The value of r, the correlation coefficient, is approximately 0.871.

(ii) There is an exact linear relation between W and L.

(iii) The line of regression of W on L has equation $W = 0.012L + 0.008$.

(iv) There is a negative correlation between the length and the height.

(v) The value of r, the correlation coefficient, is approximately 0.998.

(vi) The line of regression of W on L has equation $W = 63.5L + 16.5$.

Working:

Answers:

(a) *(i) is correct*

(b) *(iii) is correct*

Now you practise it

An easy one from November 2007 Paper 1

Tania wishes to see whether there is any correlation between a person's age and the number of objects on a tray which could be remembered after looking at them for a certain time.

She obtains the following table of results.

Age (x years)	15	21	36	40	44	55
Number of objects remembered (y)	17	20	15	16	17	12

(a) Use your graphic display calculator to find the equation of the regression line of y on x.

(b) Use your equation to estimate the number of objects remembered by a person aged 28 years.

(c) Use your graphic display calculator to find the correlation coefficient r.

(d) Comment on your value for r.

Working:

Answers:

(a) ....

(b) ....

(c) ....

(d) ....

A harder one from November 2006 Paper 2 (modified)

In an experiment a vertical spring was fixed at its upper end. It was stretched by hanging different weights on its lower end. The length of the spring was then measured. The following readings were obtained.

Load (kg) x	0	1	2	3	4	5	6	7	8
Length (cm) y	23.5	25	26.5	27	28.5	31.5	34.5	36	37.5

(a) Plot these pairs of values on a scatter diagram taking 1 cm to represent 1 kg on the horizontal axis and 1 cm to represent 2 cm on the vertical axis.

(b) Find the equation of the regression line of y on x.

(c) Draw the line of regression on the scatter diagram.

Working:

Answer:

(b) ....

OPIC 30

Chi-squared (χ^2) independence test

What do you need to know?

The chi-squared independence test is a statistical test that tells you whether or not two variables are dependent on one another to some level of accuracy. You need to know how to perform the test including all the steps, and how to interpret p values. A p value is a number that you can use to determine the validity of the test.

How do you do it?

To describe the test, it's best to go step by step through an example.

A math studies project examines the relationship between the number of push-ups a person can do in one minute and the circumference of his/her biceps. Data are collected as follows:

	Number of pushups y		
Circumference x of biceps in cm	$y < 20$	$20 \leq y < 50$	$y > 50$
$x < 28$ cm	12	18	22
$x \geq 28$ cm	19	16	14

Determine at a 95% confidence level whether or not the number of pushups is dependent on the circumference of the biceps.

Step 1: Write the null hypothesis H_0 and the alternate hypothesis H_1.

The null hypothesis is a statement that always asserts that there is no dependence between your two variables.

You always follow the form "... is not dependent on ... ."

The alternate hypothesis is the opposite: "... is dependent on ... ."

Example

H_0 = Number of pushups is not dependent on the circumference of the biceps.

H_1 = Number of pushups is dependent on the circumference of the biceps.

Step 2: Create contingency tables for observed and expected values.

The observed contingency table is just your table of data; it's usually given. The expected contingency table is a table of the same number of data values if there were absolutely no relationship at all – the number of values in each box would be in proportion to the total rather than based on some relationship.

Warning: It is not permitted to have an observed table where any of the values have a number less than 5. If you should find such a table on an exam, you will be expected to combine that row/column with another so that the new values are greater than 5 in each box.

Example

Observed table:

12	18	22
19	16	14

Expected table:

a	b	c
d	e	f

To find *a*, *b*, *c*, *d*, *e* and *f*, you need the row and column totals from the observed table:

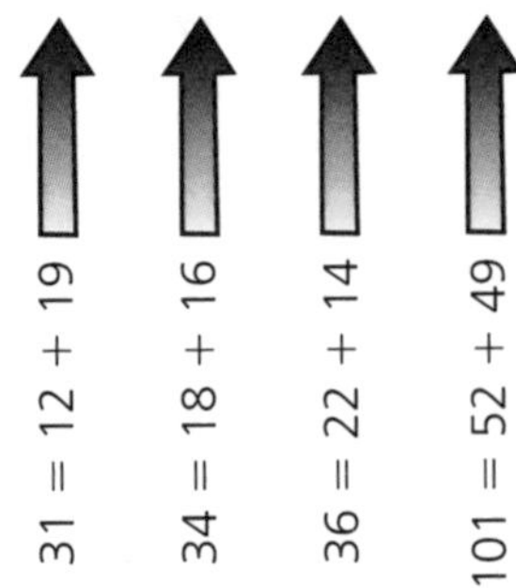

a is in the first row and first column so $a = \dfrac{52 \cdot 31}{101} = 15.96$

b is in the first row and second column, so $b = \dfrac{52 \cdot 34}{101} = 17.50$

c is in the first row and third column, so $c = \dfrac{52 \cdot 36}{101} = 18.53$

d is in the second row and first column, so $d = \dfrac{49 \cdot 31}{101} = 15.04$

e is in the second row and second column, so $e = \dfrac{49 \cdot 34}{101} = 16.50$

f is in the second row and third column, so $f = \dfrac{49 \cdot 36}{101} = 17.47$

The expected table is thus:

15.96	17.50	18.53
15.04	16.50	17.47

Step 3: Calculate the chi-squared statistic (also called the chi-squared calculated value).

There is an equation you can use from your formula booklet:

6.9 The χ^2 test statistic $\chi^2_{calc} = \sum \dfrac{(f_o - f_e)^2}{f_e}$,
where f_o are the observed frequencies, f_e are the expected frequencies

However most of the time you will do this by setting up a table as in the example below.

Example

So the chi-squared statistic is 3.39 (3sf).

Observed (*o*)	Expected (*e*)	*o* − *e*	$(o - e)^2$	$\dfrac{(o-e)^2}{e}$
12	15.96	−3.96	15.6816	0.982556
18	17.50	0.50	0.25	0.014286
22	18.53	3.47	12.0409	0.649806
19	15.04	3.96	15.6816	1.042660
16	16.50	0.50	0.25	0.015152
14	17.47	−3.47	12.0409	0.689233
			Total	3.39 (3sf)

So the chi-squared statistic is 3.39 (3s.f.)

Step 4: Calculate the "degrees of freedom".

The number of degrees of freedom (v) is found by multiplying the number of rows -1 by the number of columns -1. Or, using a formula: $v = (\text{rows} - 1) \cdot (\text{columns} - 1)$

Example

$v = (2 - 1) \cdot (3 - 1) = 1 \cdot 2 = 2$

Step 5: Find the chi-squared critical value.

Chi-squared critical values are found using the table on the last page of your formula booklet. You take the confidence or significance value that is given to you in the problem and the degrees of freedom and find the right number. Note that a 95% confidence value is equal to a 0.05 level of significance, a 98% confidence value is equal to a 0.02 level of significance, etc. Always use the confidence value with the table!

Example

From the table at the back of your formula booklet:

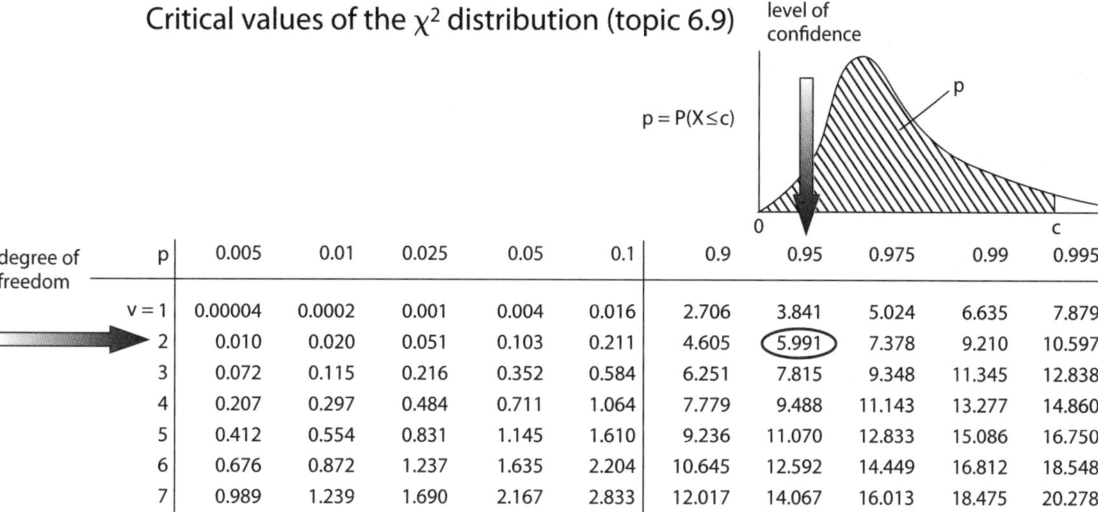

Critical values of the χ^2 distribution (topic 6.9)

degree of freedom / p	0.005	0.01	0.025	0.05	0.1	0.9	0.95	0.975	0.99	0.995
v = 1	0.00004	0.0002	0.001	0.004	0.016	2.706	3.841	5.024	6.635	7.879
2	0.010	0.020	0.051	0.103	0.211	4.605	5.991	7.378	9.210	10.597
3	0.072	0.115	0.216	0.352	0.584	6.251	7.815	9.348	11.345	12.838
4	0.207	0.297	0.484	0.711	1.064	7.779	9.488	11.143	13.277	14.860
5	0.412	0.554	0.831	1.145	1.610	9.236	11.070	12.833	15.086	16.750
6	0.676	0.872	1.237	1.635	2.204	10.645	12.592	14.449	16.812	18.548
7	0.989	1.239	1.690	2.167	2.833	12.017	14.067	16.013	18.475	20.278

So the chi-squared critical value is 5.991.

Step 6: Determine whether or not to accept the null hypothesis.

Finally we're close to the answer to the problem! To determine whether or not to accept the null hypothesis, you compare the chi-squared statistic to the chi-squared critical value. Use this key information:

> If the chi-squared statistic is LESS THAN the chi-squared critical value, you ACCEPT the null hypothesis (if $\chi^2_{calc} < \chi^2_{crit}$, then you accept H_0).
>
> If the chi-squared statistic is GREATER THAN the chi-squared critical value, you REJECT the null-hypothesis (if $\chi^2_{calc} < \chi^2_{crit}$, then you reject H_0).

SCOTT SAYS:

Students often ask me what to do if the values are equal to each other – good question. It's very unlikely that they are, but if so, you can decide to choose the next higher or lower level of confidence and accept / reject accordingly.

Example

The chi-squared statistic is 3.39 and the chi-squared critical value is 5.991, so $\chi^2_{calc} < \chi^2_{crit}$ and thus we accept the null hypothesis H_0.

Step 7: Answer the question.

Once you have determined whether or not to accept the null hypothesis, you go back to the problem and answer the question.

Example

As we have accepted the null hypothesis, the number of pushups is not dependent on the circumference of the biceps at 95% confidence.

Note about *p* values

In the new syllabus, it's expected that students know how to use "*p* values" in the chi-squared independence test. Don't worry – they make your life easier, not harder. If you are given a *p* value (or find one on your GDC), you can skip Steps 2–5 above! Now there is a new Step 6.

Step 6 when you know the *p* value:

If the *p* value is **less than** the significance value, you **reject** the null hypothesis.

If the *p* value is **greater than** the significance value, you **accept** the null hypothesis.

Don't forget that the significance value is not the same as the confidence value! If you are at 95% confidence, you have a 0.05 significance value.

Example

The *p* value for this chi-squared test is 0.184 (3sf). As the *p* value is greater than 0.05, we accept the null hypothesis.

How can my calculator help me do this?

Your calculator does **all** of this for you! Learn this carefully and you could save yourself a lot of work.

Finding the chi-squared statistic, degrees of freedom and *p* value for the example above:

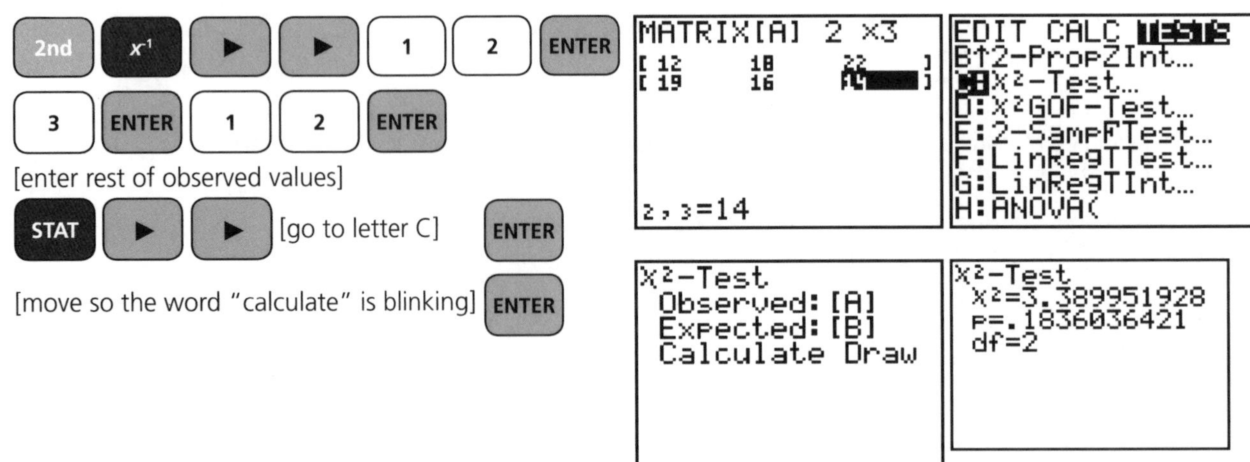

Examples from past IB papers

An easy one from November 2005 Paper 2 (modified)

In a swimming competition, the number of males and females taking part in different races is given in the table of observed values below.

	Backstroke (100 m)	Freestyle (100 m)	Butterfly (100 m)	Breaststroke (100 m)	Relay (4 × 100 m)
Male	30	90	31	29	20
Female	28	63	20	37	12

The Swimming Committee decides to perform a χ^2 test at the 5% significance level in order to test if the number of entries for the various strokes is related to gender.

(a) State the null hypothesis.

(b) Write down the number of degrees of freedom.

(c) Write down the critical value of χ^2.

Working:

(b) $(5 - 1)(2 - 1) = 4$

(c) From formula booklet: 5% significance level and 4 degrees of freedom gives 9.488 critical value.

Answers:

(a) H_0: The number of entries is not dependent on gender.

(b) 4

(c) 9.488

A harder one from May 2007 Paper 1

The local park is used for walking dogs. The sizes of the dogs are observed at different times of the day. The table below shows the numbers of dogs present, classified by size, at three different times last Sunday.

	Small	Medium	Large
Morning	9	18	21
Afternoon	11	6	13
Evening	7	8	9

(a) Write a suitable null hypothesis for a χ^2 test on this data.

(b) Write down the value of χ^2 for this data.

(c) The number of degrees of freedom is 4. Show how this value is calculated.

The critical value, at the 5% level of significance, is 9.488.

(d) What conclusion can be drawn from this test? Give a reason for your answer.

Working:

(b) Use GDC: enter matrix and do chi-squared test.

(c) (3 - 1)(3 - 1) = 4

(d) 4.33 < 9.488 so accept null hypothesis.

Answers:

(a) The size of dog is not dependent on the time of day.

(b) 4.33 (3sf)

(c) (3 - 1)(3 - 1) = 4

(d) The size of dog is not dependent on the time of day because chi-squared is less than the critical value.

Now you practise it

An easy one from November 2007 Paper 2 (modified)

Manuel conducts a survey on a random sample of 751 people to see which television programme type they watch most from the following: Drama, Comedy, Film, News. The results are as follows.

	Drama	Comedy	Film	News
Males	58	119	157	52
Females	86	98	120	61

Manuel decides to test at the 5% level of significance whether the most watched programmed type is independent of **gender**.

(a) State Manuel's null hypothesis and alternative hypothesis.

(b) Find the expected frequency for the number of females who had "Comedy" as their most-watched programmed type. Give your answer to the nearest whole number.

(c) Using your graphic display calculator, or otherwise, find the chi-squared statistic for Manuel's data.

(d) (i) State the number of degrees of freedom available for this calculation.

(ii) State the critical value for Manuel's test.

(iii) State his conclusion.

Working:

Answers:

(a) ....

(b) ....

(c) ....

(d) (i) ....

(ii) ....

(iii) ....

A harder one from May 2006 Paper 2 (modified)

For his Mathematical Studies project, Marty set out to discover if stress was related to the amount of time that students spent travelling to or from school. The results of one of his surveys are shown in the table below.

Travel time (t mins)	Number of students		
↓	high stress	moderate stress	low stress
$t \leq 15$	9	5	18
$15 < t \leq 30$	17	8	28
$30 < t$	18	6	7

He used a χ^2 test at the 5% level of significance to find out if there was any relationship between student stress and travel time.

(a) Write down the null and alternative hypotheses for this test.

(b) Write down the table of expected values. Give values to the nearest integer.

(c) Show that there are 4 degrees of freedom.

(d) Calculate the χ^2 statistic for this data.

The χ^2 critical value of 4 degrees of freedom at the 5% level of significance is 9.488.

(e) What conclusion can Marty draw from this test? Give a reason for your answer.

Working:

Answers:

(a) ....

....

(c) ....

(d) ....

(e) ....

Additional practice problems

1. From November 2007 Paper 2 *[Maximum mark: 20]*

(i) A random sample of 167 people who own mobile phones was used to collect data on the amount of time they spent per day using their phones. The results are displayed in the table below.

Time spent per day (t minutes)	$0 \le t < 15$	$15 \le t < 30$	$30 \le t < 45$	$45 \le t < 60$	$60 \le t < 75$	$75 \le t < 90$
Number of people	21	32	35	41	27	11

(a) State the modal group. *[1 mark]*

(b) Use your graphic display calculator to calculate approximate values of the mean and standard deviation of the time spent per day on these mobile phones. *[3 marks]*

(c) On graph paper, draw a fully labelled histogram to represent the data. *[4 marks]*

(ii) Manuel conducts a survey on a random sample of 751 people to see which television programme type they watch most from the following: Drama, Comedy, Film, News. The results are as follows.

	Drama	Comedy	Film	News
Males under 25	22	65	90	35
Males 25 and over	36	54	67	17
Females under 25	22	59	82	15
Females 25 and over	64	39	38	46

Manuel decides to ignore the ages and to test at the 5% level of significance whether the most watched programme type is independent of **gender.**

(a) Draw a table with 2 rows and 4 columns of data so that Manuel can perform a chi-squared test. *[3 marks]*

(b) State Manuel's null hypothesis and alternative hypothesis. *[1 mark]*

(c) Find the expected frequency for the number of females who had 'Comedy' as their most-watched programme type. Give your answer to the nearest whole number. *[2 marks]*

(d) Using your graphic display calculator, or otherwise, find the chi-squared statistic for Manuel's data. *[3 marks]*

(e) (i) State the number of degrees of freedom available for this calculation.

(ii) State the critical value for Manuel's test.

(iii) State his conclusion. *[3 marks]*

2. From November 2005 Paper 2 *[Maximum mark: 12]*

It is thought that the breaststroke time for 200 m depends on the length of the arm of the swimmer.

Eight students swim 200 m breaststroke. Their times (y) in seconds and arm lengths (x) in cm are shown in the table below.

	1	2	3	4	5	6	7	8
Length of arm, x cm	79	74	72	70	77	73	64	69
Breaststroke time, y seconds	135.1	135.7	139.3	141.0	132.8	137.0	152.9	144.0

(a) Calculate the mean and standard deviation of x and y. *[4 marks]*

(b) Given that $S_{xy} = -24.82$, calculate the correlation coefficient, r. *[2 marks]*

(c) Comment on your value for r. *[2 marks]*

(d) Calculate the equation of the regression line of y on x. *[3 marks]*

(e) Using your regression line, estimate how many seconds it will take a student with an arm length of 75 cm to swim the 200 m breaststroke. *[1 mark]*

3. From May 2004 Paper 2 *[Maximum mark: 16]*

The table below shows the number and weight (w) of fish delivered to a local fish market one morning.

Weight (kg)	**Frequency**	**Cumulative frequency**
$0.50 \le w < 0.70$	16	16
$0.70 \le w < 0.90$	37	53
$0.90 \le w < 1.10$	44	c
$1.10 \le w < 1.30$	23	120
$1.30 \le w < 1.50$	10	130

(a) (i) Write down the value of c. *[1 mark]*

(ii) On graph paper, draw the *cumulative frequency curve* for this data. Use a scale of 1 cm to represent 0.1 kg on the horizontal axis and 1 cm to represent 10 units on the vertical axis. Label the axes clearly. *[4 marks]*

(iii) Use the graph to show that the median weight of the fish is 0.95 kg. *[1 mark]*

(b) (i) The zoo buys all fish whose weights are above the 90th percentile. How many fish does the zoo buy? *[2 marks]*

(ii) A pet food company buys all the fish in the lowest quartile. What is the maximum weight of a fish bought by the company? *[3 marks]*

(c) A restaurant buys all fish whose weights are within 10% of the median weight.

(i) Calculate the minimum and maximum weights for the fish bought by the restaurant. *[2 marks]*

(ii) Use your graph to determine how many fish will be bought by the restaurant. *[3 marks]*

TOPIC 31

Differentiation and the derivative of a polynomial

What do you need to know?

The derivative of a function at a point gives you a way to calculate the slope of a tangent line at a particular point. A tangent line is a line that only touches a curve at this particular point.

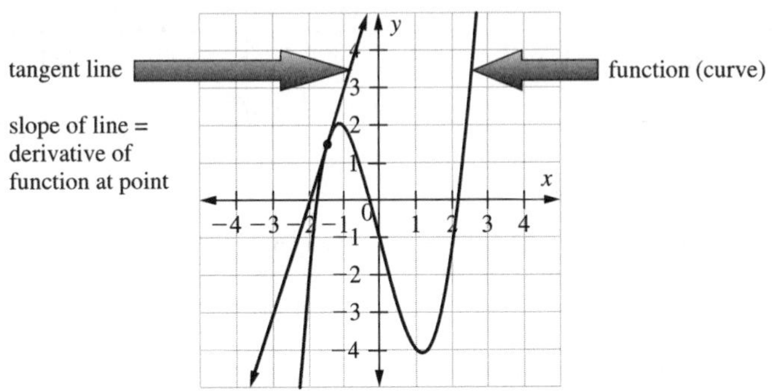

HANA AND CHRIS SAY:

If your teacher explains to you about "limits" and "derivatives by first principles", don't bother studying it for the exam. Both are not permitted to be on the exam at all!

Sometimes the IB just wants a general expression for the derivative and sometimes they want the slope of a tangent line at a specific point.

How do you do it?

To find the derivative of a function, you need to do some algebra. Fortunately the IB only asks you to find derivatives of *power series* (expressions with x, x^2, x^3, x^{-1}, x^{-2}, etc . . .). With powers, there is only one rule you need to remember: *multiply the coefficient by the exponent and reduce the exponent by one.*

multiply the 3 by 7

reduce the 7 down to 6

$$3x^7 \Rightarrow 21x^6$$

Or in maths notation: if $f(x) = 3x^7$, then $f'(x) = 21x^6$ (if they give you $y = 3x^7$, then the derivative is written $\frac{dy}{dx} = 21x^6$).

Now substitute *the x-coordinate* of the point they give you into the derivativ e to find the slope of the tangent line.

If the polynomial has negative exponents, the same rules apply! Don't forget that $x^{-1} = \frac{1}{x}$, $x^{-2} = \frac{1}{x^2}$, etc…

SCOTT SAYS:

Students often get confused about finding the derivative when there is no x. The derivative of a constant (just a number) is *always zero*. That's because if the function is a number, e.g. $f(x) = 12$, then the graph is a horizontal line and so its slope is zero.

If they ask you to find a *second derivative*, you just do the same thing again: if $f(x) = 4x^5$, then $f'(x)$ = first derivative = $20x^4$ and $f''(x)$ = second derivative = $80x^3$.

Or in the other way of writing derivatives: if $y = 4x^5$, then $\frac{dy}{dx}$ = first derivative = $20x^4$ and $\frac{d^2y}{dx^2}$ = second derivative = $80x^3$.

Examples

1. To find the slope of the tangent line of $f(x) = 3x^2 - 2x + 1$ at (1, 2), we find the derivative of the function.

   We do this by finding the derivative of each term: $3x^2$ becomes $6x$, $-2x$ becomes -2 and $+1$ becomes 0.

   So $f'(x) = 6x - 2$ and thus when we substitute $x = 1$ into this derivative, we get the slope of the tangent line: $f'(1) = 6(1) - 2 = 6 - 2 = 4$.

2. To find the derivative of the function $y = \frac{4}{x^3} + 9x$, we again do it piece by piece: $\frac{4}{x^3}$ is the same as $4x^{-3}$, so the derivative is $-12x^{-4}$, and the derivative of $9x$ is just 9.

   So if $f(x) = 4x^{-3} + 9x$, the whole derivative is $\frac{dy}{dx} = -12x^{-4} + 9$ or $\frac{dy}{dx} = \frac{-12}{x^4} + 9$.

   If we want to find the slope of the line at $x = 2$, we substitute 2 into $\frac{dy}{dx}$: $\frac{-12}{2^4} + 9 = \frac{-12}{16} + 9 = 8.25$

How can my calculator help me do this?

Your GDC can find the derivative of a function at a point very easily.

Finding the slope of the tangent line of the function $f(x) = 3x^2 - 2x + 1$ at (1, 2):

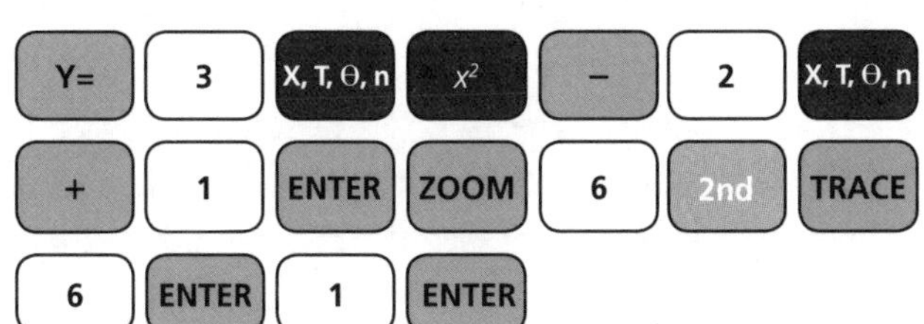

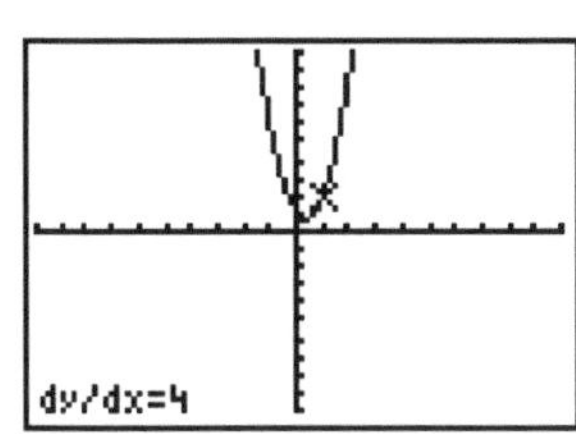

Finding the slope of the tangent line of the function $y = \frac{4}{x^3} + 9x$ at $x = 2$:

[notice that nothing appears on the graph – we need to find where the curve is by using the table and window functions]

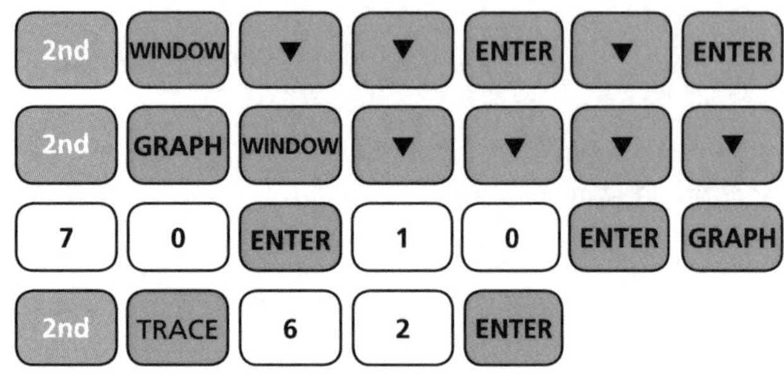

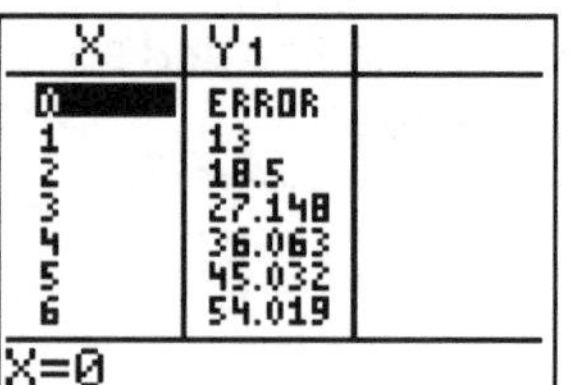

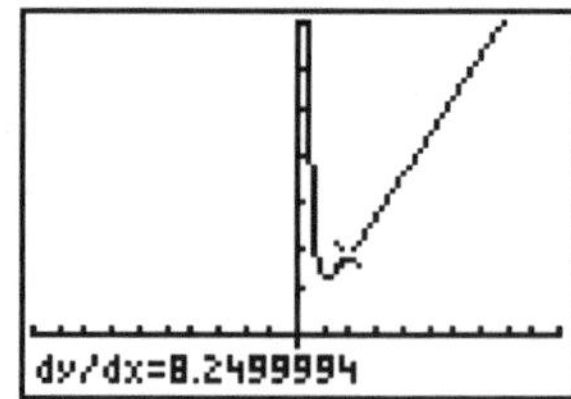

Examples from past IB papers

An easy one from May 2006 Paper 1

A function $f(x)$ is defined by

$$f(x) = 3x^4 + \frac{2}{x} - \frac{x}{4} + 1, \quad (x \neq 0).$$

(a) Calculate the 2nd derivative $f''(x)$.

(b) Find the value of $f''(x)$ at the point $\left(1, \frac{23}{4}\right)$.

Working:

(a) $f(x) = 3x^4 + 2x^{-1} - \frac{1}{4}x + 1$

$f''(x) = 12x^3 - 2x^{-2} - \frac{1}{4}$

$f'(x) = 36x^2 + 4x^{-3}$

$= 36x^2 + \frac{4}{x^3}$

(b) $f''(x)$ at $(1, \frac{23}{4}) = 36(1^2) + \frac{4}{(1^3)} = 40$

Answers:

(a) $36x^2 + \frac{4}{x^3}$

(b) 40

A harder one from May 2007 Paper 2 (modified)

A football is kicked from a point A $(a, 0)$, $0 < a < 10$ on the ground towards a goal to the right of A.

The ball follows a path that can be modelled by **part** of the graph

$$y = -0.021x^2 + 1.245x - 6.01, \ x \in \mathbb{R}, \ y \geq 0.$$

x is the horizontal distance of the ball from the origin.

y is the height above the ground.

Both x and y are measured in metres.

(a) Using your graphic display calculator or otherwise, find the value of a.

(b) Find $\frac{dy}{dx}$.

(c) (i) Use your answer to part (b) to calculate the horizontal distance the ball has travelled from A when its height is a maximum.

(ii) Find the maximum vertical height reached by the football.

Working:

(a) use GDC

(b) $\frac{dy}{dx} = 2(-0.021)x^1 + 1.245$

$= -0.042x + 1.245$

(c) (i) maximum when $f'(x) = 0$

$-0.042x + 1.245 = 0$

$x = 29.642\,857$

distance $= 29.64 - 5.30 = 24.34$

(ii) $f(29.6) = -0.021(29.6)^2 + 1.245(29.6) - 6.01$

$= 12.44$

Answers:

(a) 5.30 (3sf)

(b) $\frac{dy}{dx} = -0.042x + 1.245$

(c) (i) 24.3 m (3sf)

(ii) 12.4 m (3sf)

Now you practise it

An easy one from May 2006 Paper 2 (modified)

The function $g(x)$ is defined by

$$g(x) = \frac{1}{8}x^4 + \frac{9}{4}x^2 - 5x + 7,\ x \geq 0.$$

(a) Find $g(2)$.

(b) Calculate $g'(x)$.

Working:

Answers:

(a) ....

(b) ....

TOPIC 32

Equations of tangent lines, values of x when $f'(x)$ is given

What do you need to know?

The equation of a tangent line is the equation of a line that touches a curve at exactly one point. Its slope is the derivative of the function at that point.

Derivatives can be used to find slopes of tangent lines *and* the reverse: find the value of x for a given slope. As the slope of a tangent line at a *local maximum* or *local minimum* is zero, this is a common way for this to appear on examinations.

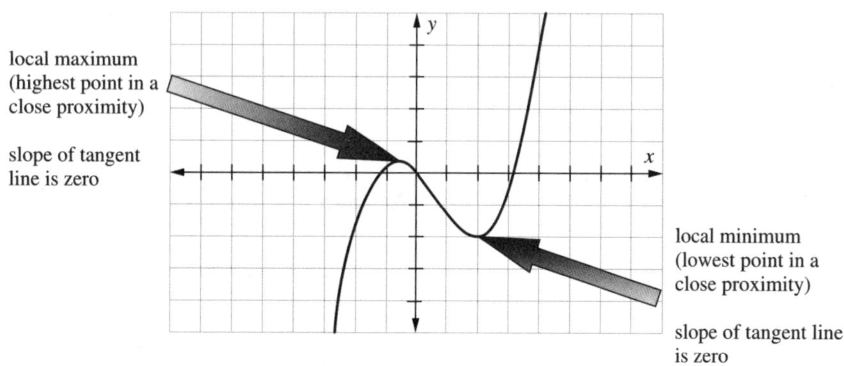

How do you do it?

To find the equation of a tangent line, first find the derivative of the function. Substitute the x-coordinate of the point into the derivative to find the slope of the tangent line (see previous section). Then substitute the x- and y-coordinates of the point into $y = mx + c$ along with the slope and solve for c.

To find the value of x given a particular slope, first find the derivative of the function (again see previous section). Then set the derivative equal to the given slope. Finally solve the equation for x.

SCOTT SAYS:

Sometimes the IB will not give you the slope of the line but will point to a local maximum or minimum, or say that the tangent line is "parallel to the x-axis". Don't panic! Just set the derivative equal to zero and solve for x.

Examples

1. If we have the function $f(x) = 2x^3 - 3x$ and want to find the equation of the tangent line at the point $(1, -1)$, first we find the derivative of f: $f'(x) = 6x^2 - 3$.

   The x-coordinate of the point is 1, so the slope of the tangent line at the point is $f'(1) = 6(1)^2 - 3 = 6 - 3 = 3$. Substituting into $y = mx + c$ gives us $-1 = 3(1) + c$. Now we can solve for c and get the equation: $-1 = 3 + c \Rightarrow c = -4 \Rightarrow y = 3x - 4$.

2. To find the coordinates of the local minimum of the equation $y = \frac{3}{x^2} + x$, we find the first derivative and set it equal to zero:

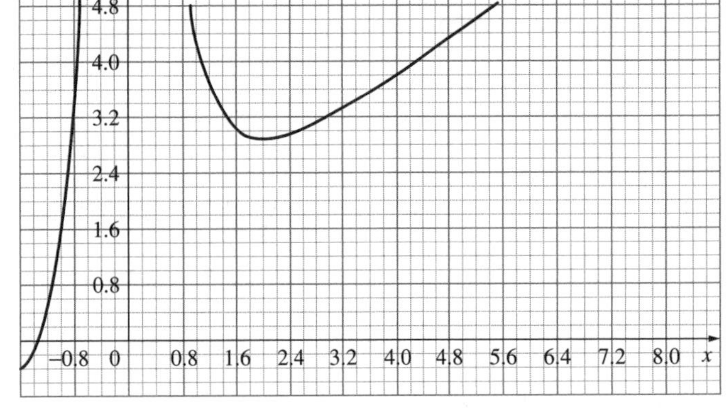

$$y = 3x^{-2} + x$$

$$\frac{dy}{dx} = -6x^{-3} + 1$$

$$-6x^{-3} + 1 = 0$$

Now we solve for x:

$$-6x^{-3} = -1$$

$$x^{-3} = \frac{1}{6}$$

$$\frac{1}{x^3} = \frac{1}{6}$$

$$x^3 = 6$$

$$x = \sqrt[3]{6} \approx 1.817$$

Lastly, we substitute 1.817 into the equation to get the y-coordinate:

$$y = \frac{3}{(1.817)^2} + 1.817 = 2.726$$

So the coordinates of the local minimum are (1.82, 2.73) (3sf).

How can my calculator help me do this?

Your GDC is extremely useful here, both in finding equations of tangent lines and finding coordinates of local maximums or local minimums.

Finding the equation of the tangent line of $f(x) = 2x^3 - 3x$ at $(1, -1)$:

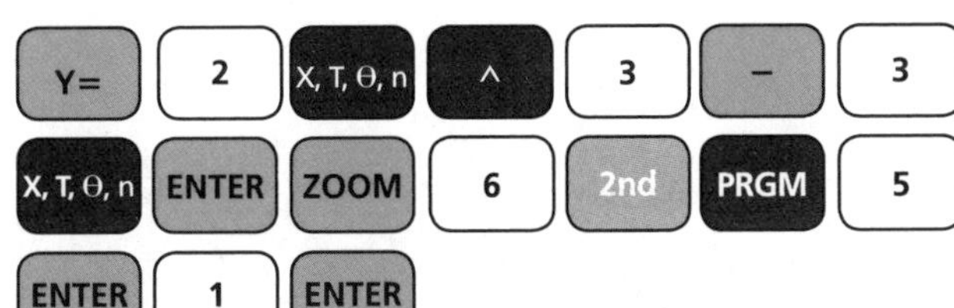

DRAW POINTS STO
1:ClrDraw
2:Line(
3:Horizontal
4:Vertical
5:Tangent(
6:DrawF
7↓Shade(

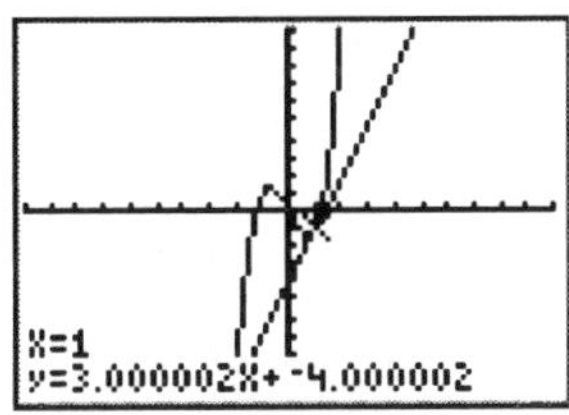

Finding the coordinates of the local minimum of $y = \frac{3}{x^2} + x$:

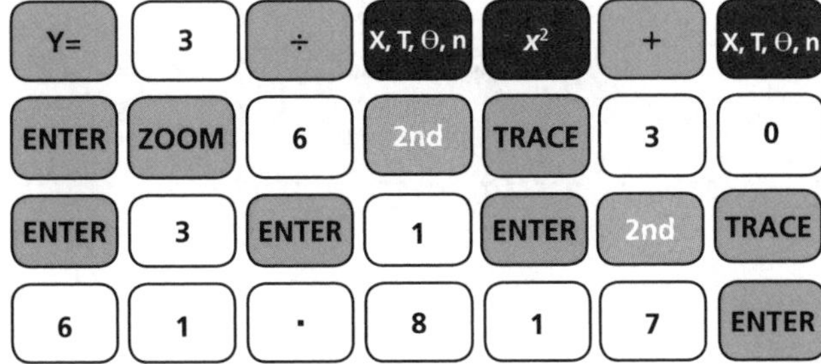

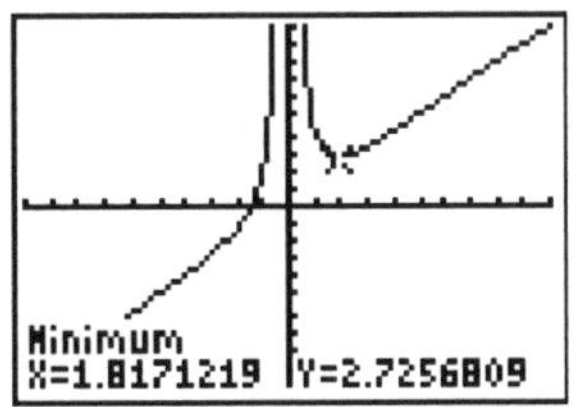

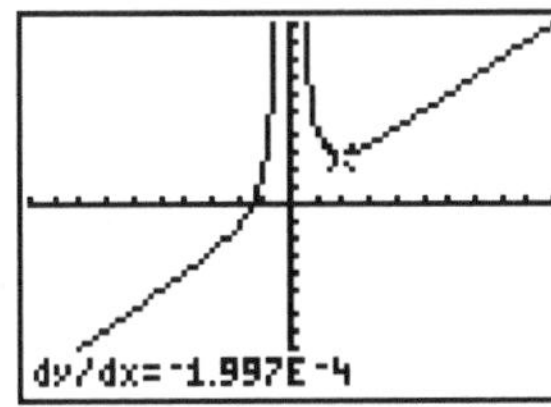

HANA AND CHRIS SAY:

Did you notice that the GDC gave a weird answer for the derivative? It said that dy/dx = −1.997E-4 instead of zero. Don't worry! −1.997E-4 is another way of writing -1.997×10^{-4} or −0.0001997. This is not zero but extremely close – good enough for us.

Examples from past IB papers

An easy one from November 2007 Paper 2

The diagram below shows the graph of a line L passing through (1, 1) and (2, 3) and the graph P of the function $f(x) = x^2 - 3x - 4$.

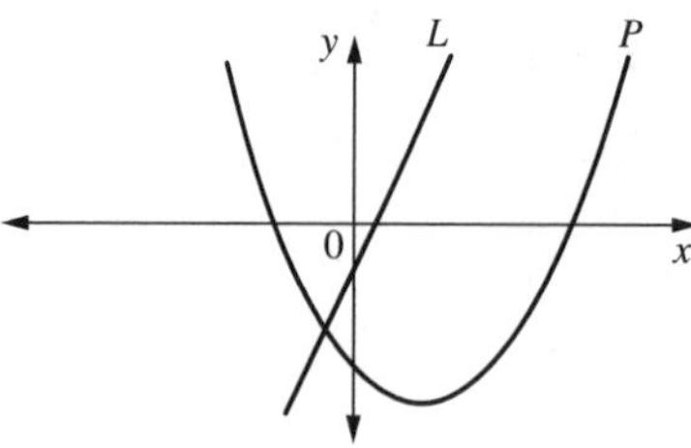

(a) Find the gradient of line L.

(b) Differentiate $f(x)$.

(c) Find the coordinates of the point where the tangent to P is parallel to the line L.

Working:

(a) $m = \frac{(3 - 1)}{(2 - 1)} = \frac{2}{1} = 2$

(b) $f(x) = x^2 - 3x - 4$

$f'(x) = 2x - 3$

(c) $2x - 3 = 2$

$x = \frac{5}{2} = 2.5$

$2.5^2 - 3(2.5) - 4 = -5.25$

Answers:

(a) $m = 2$

(b) $f'(x) = 2x - 3$

(c) $(2.5, -5.25)$

A harder one from May 2006 Paper 2 (modified)

You are given the function $g(x)$, defined by

$$g(x) = \frac{1}{8}x^4 + \frac{9}{4}x^2 - 5x + 7, \quad x \geq 0.$$

The graph of the function $y = g(x)$ has tangent T_1 at the point where $x = 2$.

(a) Show that the gradient of T_1 is 8.

(b) Find the equation of T_1. Write the equation in the form $y = mx + c$.

Working:

(a) $g'(x) = \frac{1}{2}x^3 + \frac{9}{2}x - 5$

$g'(2) = 4 + 9 - 5 = 8$

(b) $y = 8x + c$

$8 = 8(2) + c$

$-8 = c$

$y = 8x - 8$

Answers:

(a) $g'(2) = \frac{1}{2}(2^3) + \frac{9}{2}(2) - 5 = 8$

(b) $y = 8x - 8$

Now you practise it

An easy one from Specimen 2005 Paper 1 (extra question bank)

(a) Differentiate the following function with respect to x:

$$f(x) = 2x - 9 - 25x^{-1}$$

(b) Calculate the x-coordinates of the points on the curve where the gradient of the tangent to the curve is equal to 6.

Working:

Answers:

(a) ....

(b) ....

A harder one from November 2004 Paper 2

A function is given as $y = ax^2 + bx + 6$.

(a) Find $\frac{dy}{dx}$.

(b) If the gradient of this function is 2 when x is 6, write an equation in terms of a and b.

(c) If the point $(3, -15)$ lies on the graph of the function, find a second equation in terms of a and b.

Working:

Answers:

(a) ....

(b) ....

(c) ....

TOPIC 33

Increasing and decreasing functions, max/min problems (optimisation problems)

What do you need to know?

A function is considered to be *increasing* in a particular region if the slopes of the tangent lines in that region are positive.

A function is considered to be *decreasing* in a particular region if the slopes of the tangent lines in that region are negative.

If a point on a function has a perfectly horizontal tangent line, the slope of the tangent line equals zero and it is either a local maximum, a local minimum or a point of inflexion. If you're looking for the *maximum* or *minimum* quantity in some situation, it is an *optimisation* or *max/min* problem and you use calculus to solve it.

Points of inflexion are not on the IB Math Studies SL syllabus so you don't need to worry about this.

How do you do it?

To find the increasing and/or decreasing regions of a function, take the derivative of the function and then set the derivative equal to zero. Solve this equation for *x*.

This tells you where the slope(s) are equal to zero. On either side of this point there are increasing and/or decreasing regions.

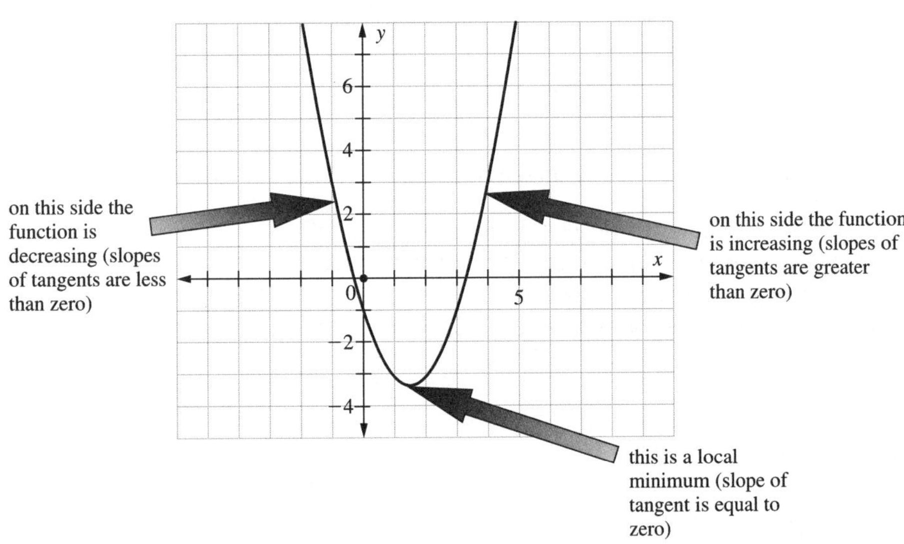

To determine whether a region is increasing or decreasing, take an *x*-coordinate from the region and substitute it into the derivative. If the derivative is positive, the region is increasing; if it is negative, the region is decreasing.

IRENE SAYS:

The IB will require you to write the region using proper notation such as $(1.5, \infty)$ or $x > 1.5$. Note that an increasing or decreasing region never includes local maximums or local minimums. You could write, "All numbers bigger than 1.5" but the maths notation is better.

To solve an optimisation or max/min problem, follow these steps:

Step 1: Define the variable.

Step 2: Derive or get a function that has only your one variable in it (you may have to do some substitution to do this).

Step 3: Take the first derivative of the function.

Step 4: Set the derivative equal to zero and solve for the variable (this solution will be your max or min).

Step 5: Solve the problem.

Examples

1. To find the increasing/decreasing regions of the curve $y = x^3 - 6x^2 + 10$, we find the derivative, set it equal to zero and solve:

   $\frac{dy}{dx} = 3x^2 - 12x$

   $3x^2 - 12x = 0$

   $3x(x - 4) = 0$

   $3x = 0 \quad \text{or} \quad x - 4 = 0$

   $x = 0 \quad \text{or} \quad x = 4$

   so there are three regions: one less than 0, one between 0 and 4, and another greater than 4. Testing a point from each region tells us whether it is increasing or decreasing:

testing $x = -1$:	testing $x = 2$:	testing $x = 7$:
$3(-1)^2 - 12(-1)$	$3(2)^2 - 12(2)$	$3(7)^2 - 12(7)$
$= 3 + 12 = 15$	$= 12 - 24 = -12$	$= 147 - 84 = 63$
$\Rightarrow$ increasing	$\Rightarrow$ decreasing	$\Rightarrow$ increasing

2. The student council is planning a school dance in the gymnasium and has to define the dance floor so people know where to boogie. One side of the dance floor is against the wall and the other three sides will be roped off. If the council has 20 metres of rope, what should the dimensions of the dance floor be in order to maximise the area?

   x = width of dance floor, $20 - 2x$ = length of dance floor

   Area $= A = x(20 - 2x) = 20x - 2x^2$

   $\frac{dA}{dx} = 20 - 4x$

   $0 = 20 - 4x$

   $4x = 20$

   $x = 5$

   The width of the dance floor should be 5 m and the length should be $20 - 2(5) = 10$ m.

How can my calculator help me do this?

The GDC can find local maximums and minimums (see previous section) but can also test regions for increasing or decreasing. It's usually easy to see these regions by just looking at a graph!

Finding the increasing and decreasing regions for $y = x^3 - 6x^2 + 10$:

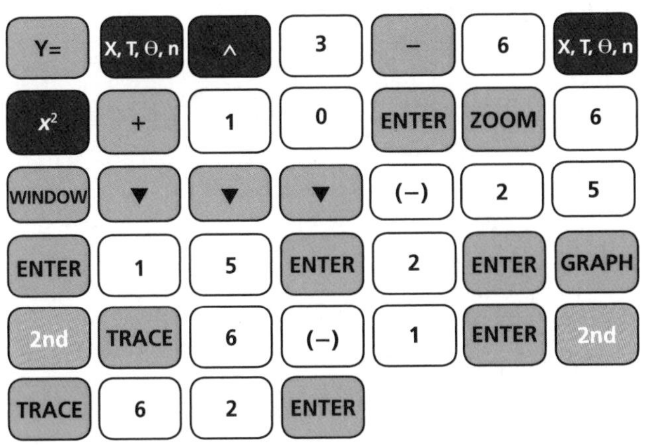

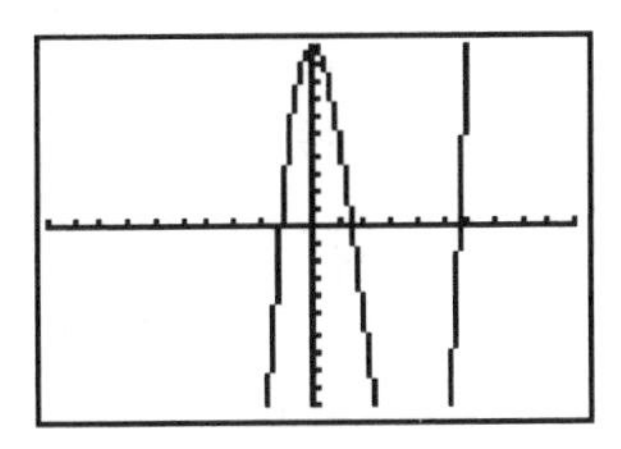

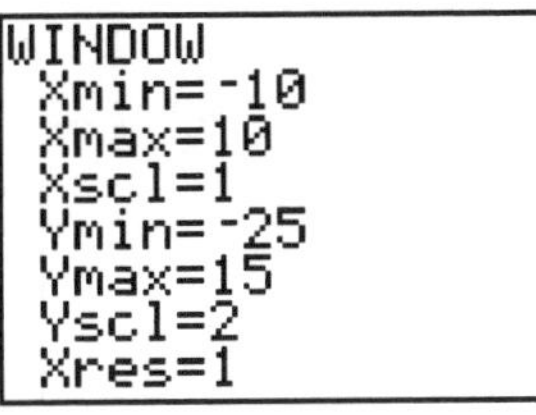

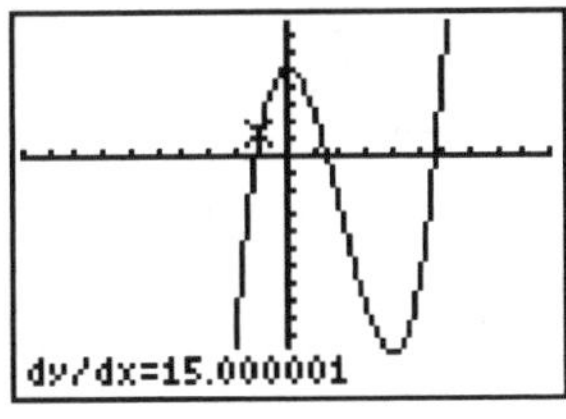

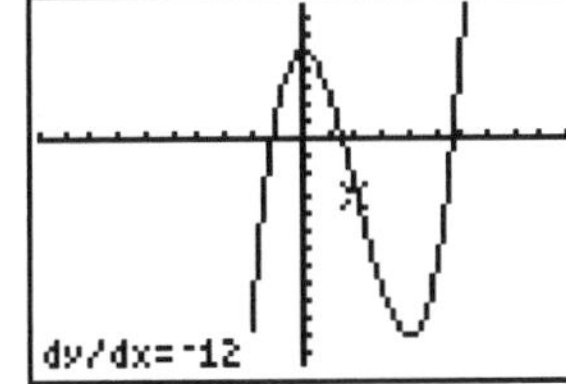

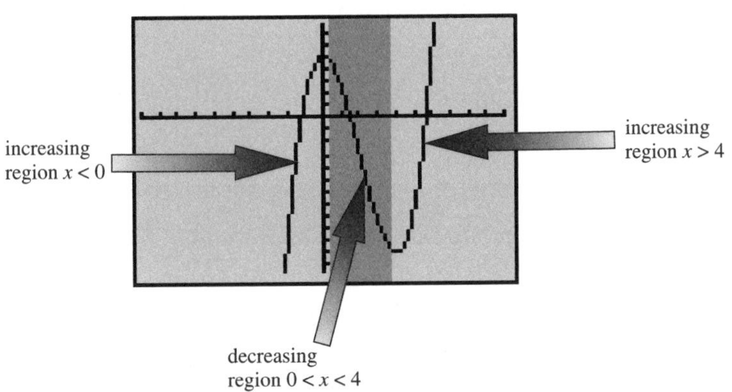

Examples from past IB papers

An easy one from May 2005 Paper 2 (modified)

The function $f(x)$ is given by the formula $f(x) = 2x^3 - 5x^2 + 7x - 1$.

(a) Calculate $f'(x)$.

(b) Evaluate $f'(2)$.

(c) State whether the function $f(x)$ is increasing or decreasing at $x = 2$.

Working:

(a) $f'(x) = 3(2)x^{3-1} - 2(5)x^{2-1} + 7x^{1-1}$

$= 6x^2 - 10x + 7$

(b) $f'(2) = 6(2^2) - 10(2) + 7$

$= 24 - 20 + 7$

$= 11$

(c) $f'(2) = 11$ which is positive, so $f(x)$ is increasing at $x = 2$.

Answers:

(a) $f'(x) = 6x^2 - 10x + 7$

(b) $f'(2) = 11$

(c) increasing

A harder one from November 2006 Paper 2 (modified)

A farmer has a rectangular enclosure with a straight hedge running down one side. The area of the enclosure is 162 m². He encloses this area using x metres of the hedge on one side as shown on the diagram below.

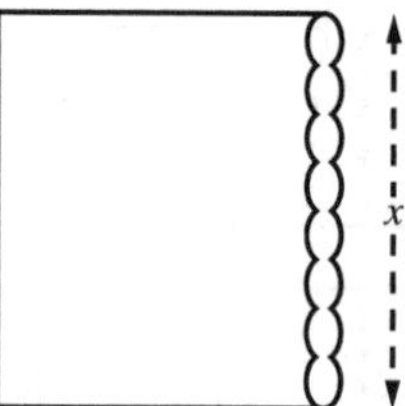

diagram not to scale

(a) If he uses y metres of fencing to complete the enclosure, show that $y = x + \frac{324}{x}$.

The farmer wishes to use the least amount of fencing.

(b) Find $\frac{dy}{dx}$.

(c) Find the value of x which makes y a minimum.

(d) Calculate this minimum value of y.

Working:

(a) $y = x + z + z$

$xz = 162$

$z = \frac{162}{x}$

$y = x + \frac{2(162)}{x}$

$y = x + \frac{324}{x}$

(b) $\frac{dy}{dx} = 1x^0 + \frac{324}{(-1)x^{-2}}$

$= 1 - \frac{324}{x^2}$

(c) $\frac{dy}{dx} = 0$

$1 - \frac{324}{x^2} = 0$

$x^2 = 324$

$x = 18$

(d) $y = 18 + \frac{324}{18}$

$= 36$

Answers:

(a) see working

(b) $\frac{dy}{dx} = 1 - \frac{324}{x^2}$

(c) $x = 18$ m

(d) $y = 36$ m

Now you practise it

An easy one from November 2004 Paper 2 (modified)

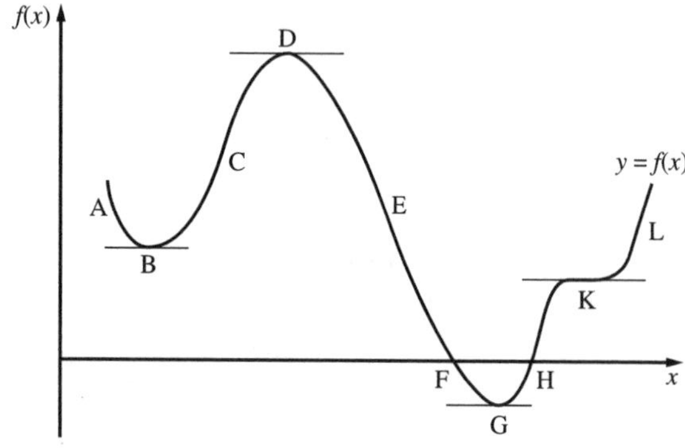

Given the graph of $f(x)$ state

(a) the intervals from A to L in which $f(x)$ is increasing

(b) the intervals from A to L in which $f(x)$ is decreasing

(c) a point that is a maximum value

(d) a point that is a minimum value.

Working:

Answers:

(a) ....

(b) ....

(c) ....

(d) ....

A harder one from November 2001 Paper 2 (modified)

A rectangular piece of card measures 24 cm by 9 cm. Equal squares of length x cm are cut from each corner of the card as shown in the diagram below. What is left is then folded to make an **open** box, of length l cm and width w cm.

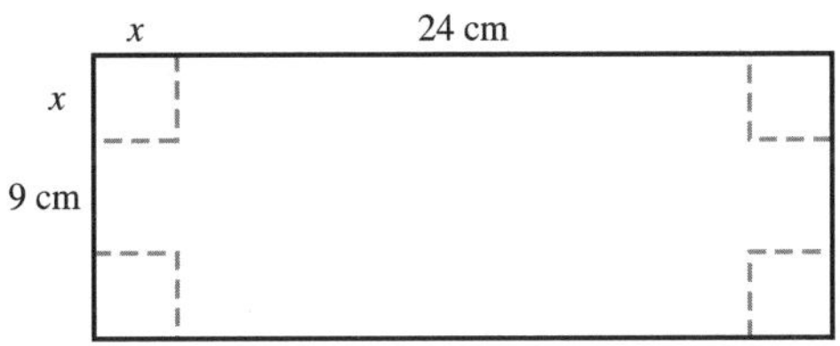

(a) Given that $l = 24 - 2x$ and $w = 9 - 2x$, show that the volume (B m^3) of the box is given by $B = 4x^3 - 66x^2 + 216x$.

(b) Find $\frac{dB}{dx}$.

(c) (i) Find the value of x which gives the maximum volume of the box.

(ii) Calculate the maximum volume of the box.

Working:

Answers:

(a) ....

(b) ....

(c) (i) ....

(ii) ....

Additional practice problems

1. From November 2004 Paper 2 *[Maximum mark: 15]*

(i) Consider the function $g(x) = x^4 + 3x^3 + 2x^2 + x + 4$.

Find

(a) $g'(x)$ *[3 marks]*

(b) $g'(1)$ *[2 marks]*

(ii) The cost of producing a mathematics textbook is \$15 (US dollars) and it is then sold for \$$x$.

(a) Find an expression for the profit made on each book sold. *[1 mark]*

A total of $(100\,000 - 4000x)$ books is sold.

(b) Show that the profit made on all the books sold is $P = 160\,000x - 4000x^2 - 1\,500\,000$. *[3 marks]*

(c) (i) Find $\frac{dP}{dx}$. *[2 marks]*

(ii) Hence calculate the value of x to make a maximum profit. *[2 marks]*

(d) Calculate the number of books sold to make this maximum profit. *[2 marks]*

2. From May 2004 Paper 2 *[Maximum mark: 14]*

The height (cm) of a daffodil above the ground is given by the function $h(w) = 24w - 2.4w^2$, where w is the time in weeks after the plant has broken through the surface ($w \geq 0$).

(a) Calculate the height of the daffodil after two weeks. *[2 marks]*

(b) (i) Find the rate of growth, $\frac{dh}{dw}$. *[2 marks]*

(ii) The rate of growth when $w = k$ is 7.2 cm per week. Find k. *[3 marks]*

(iii) When will the daffodil reach its maximum height? What height will it reach? *[4 marks]*

(c) Once the daffodil has reached its maximum height, it begins to fall back towards the ground. Show that it will touch the ground after 70 days. *[3 marks]*

3. From May 2003 Paper 3 *[Maximum mark: 20]*

Consider the function $f(x) = x^3 - 4x^2 - 3x + 18$.

(a) (i) Find $f'(x)$.

(ii) Find the coordinates of the maximum and minimum points of the function. *[10 marks]*

(b) Find the values of $f(x)$ for a and b in the table below: *[2 marks]*

(c) Using a scale of 1 cm for each unit on the x-axis and 1 cm for each 5 units on the y-axis, draw the graph of $f(x)$ for $-3 \leq x \leq 5$. Label your graph clearly. *[5 marks]*

x	−3	−2	−1	0	1	2	3	4	5
$f(x)$	−36	a	16	b	12	4	0	6	28

(d) The gradient of the curve at any particular point varies. Within the interval $-3 \leq x \leq 5$, state all the intervals where the gradient of the curve at any particular point is

(i) negative;

(ii) positive. *[3 marks]*

TOPIC 34

Currency conversions

What do you need to know?

Currency is converted from one kind of money to another using *exchange rates* with or without fees. Exchange rates can be one simple ratio (e.g. US\$1.00 = ¥101.86) or there may be an exchange table (e.g. an exchange bureau in London buying US dollars at \$1.90 per pound and selling US dollars at \$1.80 per pound).

Currency exchange fees come in several forms. Sometimes there is a flat fee in local currency, sometimes the fee is a percentage of the money being exchanged, and sometimes it is a combination of both.

HANA SAYS:

When doing these problems it really helps if you know what the main world currencies are and roughly what they are worth! For example, 1 euro is worth a bit less than a British pound and about $1\frac{1}{2}$ times more than a US dollar; US dollars are worth about the same as Canadian dollars and both are worth a bit more than Australian dollars.

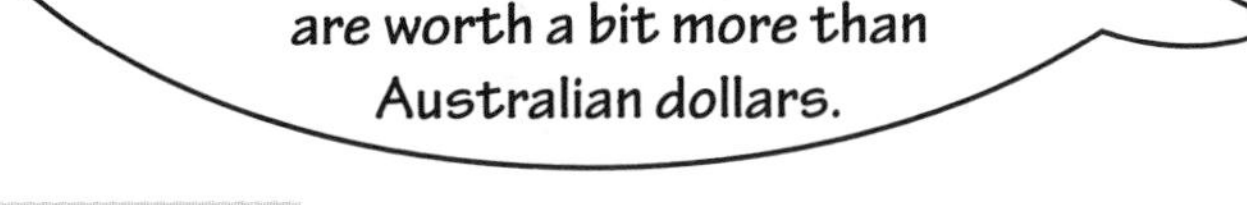

CHRIS SAYS:

Look up rates on the internet at http://finance.yahoo.com/currency. You should know at least US dollars, Australian dollars, Canadian dollars, UK pounds, Japanese yen, euros and Swiss francs.

How do you do it?

To convert currency using a ratio, set up an equation and solve it using cross multiplication:

$$\frac{\text{exchange rate of 1st currency}}{\text{exchange rate of 2nd currency}} = \frac{\text{amount you want of 1st currency}}{\text{amount you want of 2nd currency}}$$

If you are given a buy–sell table, it is often confusing which number to use. The main idea is that it's the *bureau* who is buying or selling the currency, not you. Still confused? Do the calculation twice, once with each rate, and use the one which is worse for the person!

Fee calculations are straightforward: subtract the flat fee or the percentage from either the initial amount or the answer (whichever one is easier to deal with).

Examples

1. The exchange rate between South African rand (ZAR) and Australian dollars (AUD) is 1 AUD = 6.6128 ZAR. If we want to convert 2500 ZAR to AUD, we can use the ratio:

$$\frac{1\text{ AUD}}{6.6128\text{ ZAR}} = \frac{x\text{ AUD}}{2500\text{ ZAR}}$$

$$6.6128x = 2500$$

$$x = 378.05\text{ AUD}$$

Note that we could have just done 2500 ÷ 6.6128 to get the answer, but if you use the ratio you will never be confused about whether to multiply or divide!

2. A bank in the Philippines is buying and selling foreign currency at the following rates for one Philippine peso (PHP):

	Buy	Sell
Euro (€)	0.0149	0.0132
Canadian dollar (C$)	0.0232	0.0221
Japanese yen (¥)	2.3747	2.2422
British pound (£)	0.0120	0.0105

If there is a 10 PHP + 3% commission for each currency exchange and we want to change ¥2000 into British pounds, we first need to convert into pesos, deduct the commission and then convert to British pounds:

$$\frac{1\text{ PHP}}{¥2.3747} = \frac{x}{2000}$$

Converting ¥2000 to pesos: $2.3747x = 2000$

$$x = 842.2116478\text{ PHP}$$

Now deduct the commission:

$842.216478 \times 0.03 = 25.2663494$

$25.2663494 + 10 = 35.2663494$ PHP total commission

$842.2116478 - 35.2663494 = 806.9452983$ PHP left after commission

Finally convert back to British pounds:

$$\frac{1\text{ PHP}}{£0.0105} = \frac{806.9452983}{x}$$

$$x = 0.0105 \times 806.9452983$$

$$x = £8.47$$

IRENE SAYS:

The IB will give you the level of accuracy for all currency questions. This is good and bad! Good because you know how much to round the answer to; bad because there is now a special finance penalty if you round a currency answer incorrectly! Make sure you *follow instructions* and *don't round until the end.*

How can my calculator help me do this?

Your calculator is not much help here besides doing the arithmetic for you. One useful feature, though, is to store the answer to each calculation in your calculator, unrounded, until you are done.

You can store a number by typing **STO➤** **ALPHA** **MATH**, and you can retrieve the stored number by typing **ALPHA** **MATH** **ENTER**. If you already have a number stored in A, you can use B, C, D, etc!

```
2000/2.3747
           842.2116478
Ans→A
           842.2116478
A
           842.2116478
```

Examples from past IB papers

An easy one from May 2006 Paper 1

Sven is travelling to Europe. He withdraws \$800 from his savings and converts it to Euros. The local bank is buying Euros at \$1 : €0.785 and selling Euros at \$1 : €0.766.

(a) Use the appropriate rate above to calculate the amount of Euros Sven will receive.

(b) Suppose the trip is cancelled. How much will he receive if the Euros in part (a) are changed back to dollars?

(c) How much has Sven lost after the two transactions? Express your answer as a percentage of Sven's original \$800.

Working:

(a) $800 \times 0.766 = 612.80$

(b) $\frac{612.80}{0.785} = 780.64$

(c) $\left(\frac{800 - 780.64}{800}\right) \times 100 = 2.42\%$

Answers:

(a) €612.80

(b) \$780.64

(c) 2.42%

A harder one from November 2003 Paper 1

A Swiss bank shows currency conversion rates in a table. Part of the table is shown below, which gives the exchange rate between British pounds (GBP), US dollars (USD) and Swiss francs (CHF).

	Buy	Sell
GBP	2.3400	2.4700
USD	1.6900	1.7700

This means that the bank will **sell** its British pounds to a client at an exchange rate of 1 GBP = 2.4700 CHF.

(a) What will be the selling price for 1 USD?

Andrew is going to travel from Europe to the USA. He plans to exchange 1000 CHF into dollars. The bank sells him the dollars and charges 2% commission.

(b) How many dollars will he receive? Give your answer to the nearest dollar.

Working:

(b) 2% of 1000 CHF = 20 CHF

Amount = 1000 − 20

= 980 CHF to exchange

$\frac{980}{1.7700}$ USD

= 553.67 USD

Answers:

(a) 1.7700 CHF

(b) 554 USD

Now you practise it

An easy one from May 2003 Paper 1

Zog from the planet Mars wants to change some Martian Dollars (MD) into US Dollars (USD). The exchange rate is 1 MD = 0.412 USD. The bank charges 2% commission.

(a) How many US Dollars will Zog receive if she pays 3500 MD?

Zog meets Zania from Venus where the currency is Venusian Rupees (VR). They want to exchange money and avoid bank charges. The exchange rate is 1 MD = 1.63 VR.

(b) How many Martian Dollars, to the nearest dollar, will Zania receive if she gives Zog 2100 VR?

Working:

Answers:

(a) ....

(b) ....

A harder one from November 2002 Paper 1

Frederick had to change British pounds (GBP) into Swiss francs (CHF) in a bank. The exchange rate is 1 GBP = 2.5 CHF. There is also a bank charge of 3 GBP for each transaction.

(a) How many Swiss francs would Frederick buy with 133 GBP?

(b) Let s be the number of Swiss francs received in exchange for b GBP. Express s in terms of b.

(c) Frederick received 430 CHF. How many British pounds did he exchange?

Working:

Answers:

(a) ....

(b) ....

(c) ....

TOPIC 35 Simple and compound interest

What do you need to know?

Interest is money you earn by lending money to someone (usually a bank) for a period of time. In general, there are two kinds of interest:

- simple interest: interest you earn *at the end* of the time period
- compound interest: interest you earn *throughout* the time period.

Compound interest will always give you more money than simple interest (assuming everything else is the same!) because with compound interest, you are earning interest on the original money *and* the interest you have already earned. You make money on money!

Interest is usually expressed as a percentage per annum (per year), though sometimes it can be stated per month, per quarter, etc.

How do you do it?

To calculate simple interest, you multiply the interest rate by the amount you started with (the capital) and by the number of time periods, and then add that to your original amount. There is a nice formula in your booklet you can use:

Remember that r is the interest *not converted to decimal form*! For example, if the interest rate is 7%, $r = 7$, not 0.07. Also, this formula gives you the *interest* you earn, not the *final amount in your account.*

8.2 Simple interest $I = \frac{Crn}{100}$, where C = capital, $r\%$ = interest rate, n = number of time periods, I = interest

To calculate compound interest, there is again a formula in your booklet but it is much messier!

8.3 Compound interest $I = C\left(1 + \frac{r}{100}\right)^n - C$, where C = capital, $r\%$ = interest rate, n = number of time periods, I = interest,
which is the special case $k = 1$ of $I = C\left(1 + \frac{r}{100k}\right)^{kn} - C$,
where C = capital, $r\%$ = nominal interest rate, n = number of time periods, k = number of compounding periods in a single year and I = interest

Basically if the compounding is annual, you use the first formula and if the compounding is more frequent than that, you use the second.

Annual compounding: use the first formula – substitute the interest rate into r, calculate the interest I, and then answer the question

Quarterly compounding: use the second formula – substitute the interest rate into r, put $k = 4$ in both places, calculate the interest I, and then answer the question

Monthly compounding: use the second formula – substitute the interest rate into r, put $k = 12$ in both places, calculate the interest I, and then answer the question

Weekly, daily and other compounding are similar to quarterly and monthly.

Examples

1. You invest £14 000 into a bank account that is offering 3.25% per annum simple interest. After 12 years you would earn

$$I = \frac{£14000 \times 3.25 \times 12}{100} = £5460$$

in interest. So after 12 years there will be a total of £19 460 in the account.

2. You invest \$500 into a bank account that pays 4.5% per annum compounded annually. After 8 years you would earn

$$I = \$500\left(1 + \frac{4.5}{100}\right)^8 - \$500 = \$211.05 \text{ (2dp)}$$

in interest. So after 8 years there will be a total of \$711.05 in the account.

SCOTT SAYS:

Notice how silly it was to subtract the \$500 and then add it back on a minute later! If the question asks for the final amount, rather than just the interest, don't subtract *C* in the original formula and you'll get the final amount straight away.

3. Your Nigerian grandmother gives you ₦300 000 as a gift for earning 43 points on your IB diploma! You decide to invest it in a bank that is offering 6.2% interest per annum, compounded daily, and save it for 3 years. After the 3 years, you will have this much in the account:

$$A = 300000\left(1 + \frac{6.2}{100 \cdot 365}\right)^{365 \cdot 3} = ₦361\ 320.97 \text{ (2dp)}$$

How can my calculator help me do this?

There is a *fabulous* way to do compound interest on your GDC if you are a bit adventurous. You use the *finance app* which is actually allowed on all IB exams (your IB coordinator cannot delete it; the app is built into the calculator). If you can remember how to use the finance app, it's really useful!

Finding the final amount after investing \$500 at 4.5% per annum compounded annually for 8 years:

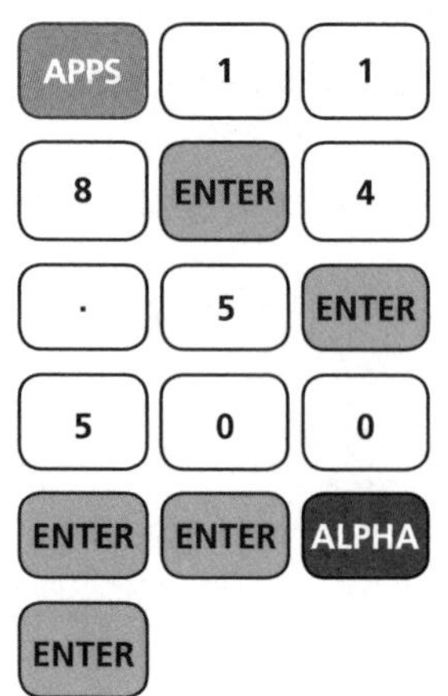

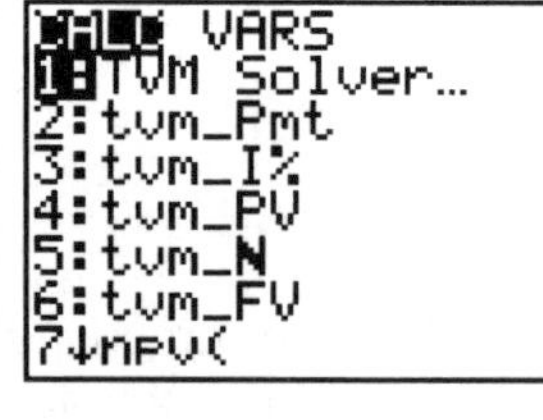

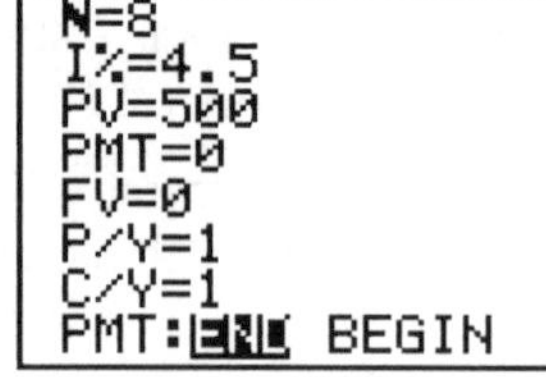

N=8
I%=4.5
PV=500
PMT=0
▪FV=-711.0503064
P/Y=1
C/Y=1
PMT:END BEGIN

Finding the final amount after investing ₦300 000 at 6.2% per annum compounded daily for 3 years:

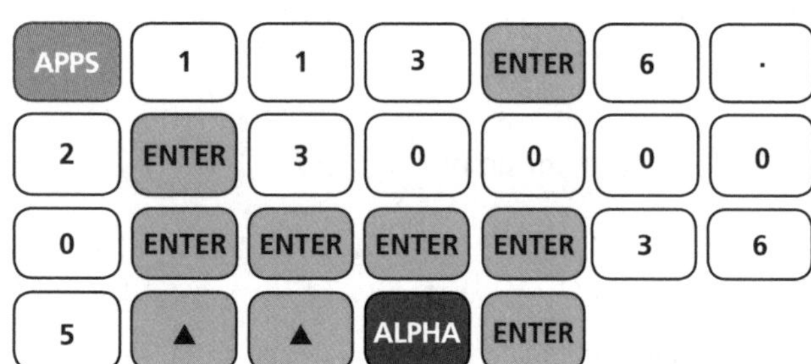

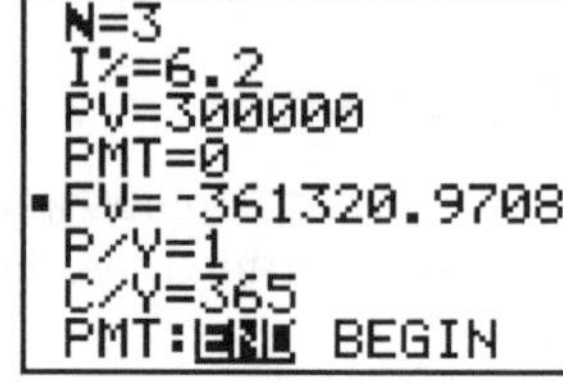

Examples from past IB papers

An easy one from November 2004 Paper 1

Bobby is spending a year travelling from America to France and Britain. Consider the following exchange rates.

1 US dollar (USD) = 0.983 Euros

1 British Pound (GBP) = 1.59 Euros

(a) Bobby changes 500 USD into Euros.

(i) Calculate how many Euros he receives.

He spends 328 Euros in France and changes the remainder into GBP.

(ii) Calculate how many GBP he receives.

While in Britain Bobby decides to put this money in a bank that pays 6% simple interest per annum, and he gets a part-time job to cover his expenses. Bobby remains in Britain for six months.

(b) Calculate how much interest he receives for the six months.

Working:

(a) (i) $500 \times 0.983 = 491.50$ euros

(ii) $491.50 - 328 = 163.50$ euros

$\frac{163.50}{1.59} = 102.83$ GBP

(b) $\frac{(102.83 \times 0.5 \times 6)}{100}$

$= 3.08$ GBP

Answers:

(a) (i) 491.50 euros

(ii) 102.83 GBP

(b) 3.08 GBP

A harder one from November 2003 Paper 1

David invests 6000 Australian dollars (AUD) in a bank offering 6% interest compounded annually.

(a) Calculate the amount of money he has after 10 years.

(b) David then withdraws 5000 AUD to invest in another bank offering 8% interest compounded annually. Calculate the **total** amount he will have in both banks at the end of one more year. Give your answer correct to the nearest Australian dollar.

Working:

(a) $6000\,(1.06)^{10}$

$= 10\,745$ AUD

(b) $10\,745 - 5000$

$= 5745$

$5000 \times 1.08 + 5745 \times 1.06$

$= 11\,489.70$ AUD

Answers:

(a) 10 745 AUD

(b) 11 490 AUD

Now you practise it

An easy one from May 2000 Paper 1

Hassan invested 10 000 CHF at the end of 1971. The interest was 5% per annum. How much interest **in total** would Hassan have earned at the end of the year 1999 if

(a) he had removed the interest from his account at the end of each year;

(b) he had **not** removed the interest from his account at the end of each year?

Working:

Answers:

(a) ....

(b) ....

A harder one from November 2000 Paper 1

John invests XUSD in a bank. The bank's stated rate of interest is 6% per annum, compounded **monthly**.

(a) Write down, in terms of X, an expression for the value of John's investment after one year.

(b) What rate of interest, when compounded **annually** (instead of monthly) will give the same value of John's investment as in part (a)? Give your answer correct to three significant figures.

Working:

Answers:

(a) ....

(b) ....

TOPIC 36

Loan and savings tables

What do you need to know?

Loan and savings tables are used to calculate, usually on a monthly basis, how much you have in an account after making regular payments. This is different from compound interest in that you are not putting money all in once and leaving it there for a long time; you are putting money into an account or paying back a loan every month/week/year/etc.

How do you do it?

There is no formula for these problems (a formula does exist, of course, but it's pretty complicated). Instead, you are expected to create a *loan table* or a *savings table*. To do this, you need to keep careful track of how much you start with, what your payments are, what interest you have earned/paid, and how much is now in the account. The tables vary in what they look like, but in general they take this form:

Loan table for a €4000 loan with monthly payments of €100 each at 12% interest per annum paid monthly. Savings tables work the same way but you're adding money, not taking it away.

STEP (1) 12% per annum = 1% per month, so the interest paid this month is €4000 × 0.01 = €40

STEP (2) After you have paid the interest, you can pay off the loan from what is left: €100 − €40 = €60

Date	Balance	Payment	Interest	Principal paid
Jan 1	€4000.00			
Feb 1	€3940.00	€100.00	€40.00	€60.00
Mar 1	€3879.40	€100.00	€39.40	€60.60
Apr 1	€3818.20	€100.00	€38.80	€61.20

STEP (3) Now deduct your payment from your old balance to get your new balance: €4000 − €60 = €3940

STEP (1) 12% per annum = 1% per month, so the interest paid this month is €4000 × 0.01 = €40

STEP (2) Now you add this interest to your payment: €100 + €40 = €140

Date	Balance	Payment	Interest	Principal paid
Jan 1	€4000.00			
Feb 1	€4140.00	€100.00	€40.00	€140.00
Mar 1	€4281.40	€100.00	€41.40	€141.40
Apr 1	€4424.21	€100.00	€42.81	€142.81

STEP (3) Your new balance found from adding your new money to your old balance: €4000 + €140 = €4140

IRENE SAYS:

The IB sets up these tables in many different ways. Don't memorise this exact table! Figure out how it works so you can apply it to whatever table happens to appear on the exam!

Examples

1. Irene takes out a ¥20,000 loan to pay for all the red pens she is using to mark IB examinations. She agrees to pay ¥1000 monthly payments at 8% interest per annum compounded monthly. To find out how much she would owe after four payments, she creates a loan table, rounding values to the nearest Yen:

Month	Balance	Payment	Interest	Principal
	¥20 000			
1	¥19 133	¥1000	¥133	¥867
2	¥18 261	¥1000	¥128	¥872
3	¥17 383	¥1000	¥122	¥878
4	¥16 499	¥1000	¥116	¥884

The interest this time is calculated by taking 8% and dividing by 12 (to convert to months): 0.08 ÷ 12 = 0.006 667 or 0.6667% per month.

So the first interest payment would be ¥20 000 × 0.006 667 = ¥133.

This leaves ¥1000 – ¥133 = ¥867 left in principal to pay back the loan this month.

Thus the new balance is ¥20 000 – ¥867 = ¥19 133.

The rest just continues in the same way until you get to the final answer at the end of month 4: ¥16 499 left!

2. Your generous Nigerian grandmother (see previous section) now decides to give you the same ₦300 000 and an additional ₦20 000 each week for the next eight weeks. You wisely tell grandmother to deposit the money directly into your bank which is paying you 4% per annum. If you want to find out how many naira you will have at the end of the eight weeks, you can use a savings table:

Week	Balance	Payment	Interest	Principal
	₦300 000.00			
1	₦320 230.77	₦20 000	₦230.77	₦20 230.77
2	₦340 477.10	₦20 000	₦246.33	₦20 246.33
etc...				
8	₦462 282.56	₦20 000	₦339.96	₦20 339.96

The interest this time is calculated by taking 4% and dividing by 52 (to convert to weeks): 0.04 ÷ 52 = 0.000 769 23 or 0.076 923% per month.

So the first interest payment would be ₦300 000 × 0.000 769 23 = ₦230.77.

This yields ₦20 000 + ₦230.77 = ₦20 230.77 to add to your account.

Thus the new balance is ₦300 000 + ₦20 230.77 = ₦320 230.77.

The rest just continues in the same way until you get to the final answer at the end of week 8: ₦462 282.56.

How can my calculator help me do this?

Again, the finance app is what to use here!

Finding the final balance after taking out a ¥20 000 loan with ¥1000 monthly payments at 8% interest per annum compounded monthly for 4 payments:

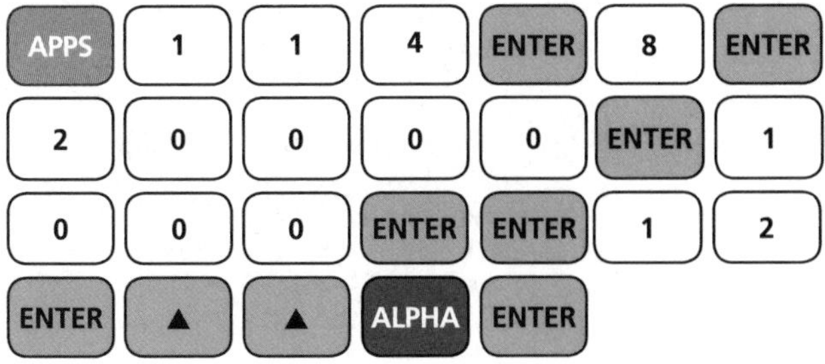

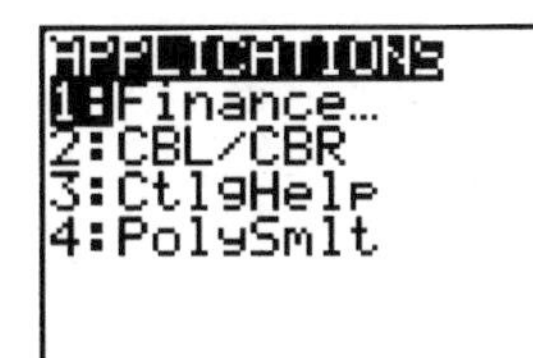

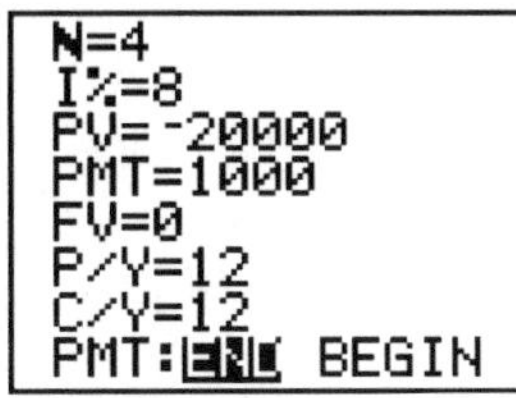

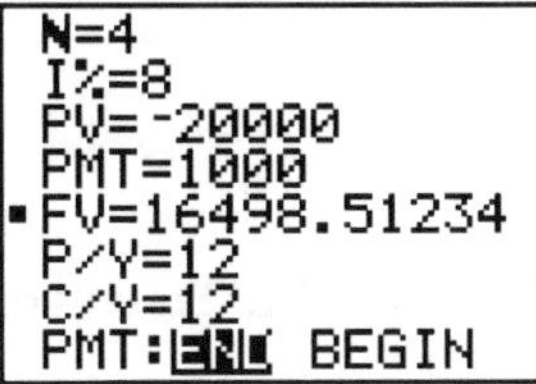

Finding the final balance after investing ₦300 000 plus ₦20 000 per week for the next eight weeks at 4% per annum:

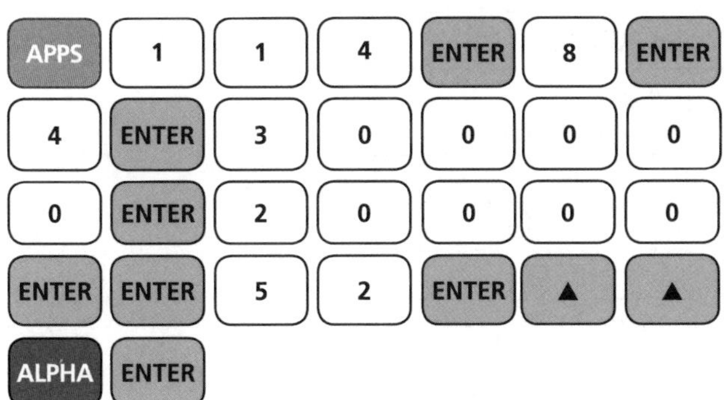

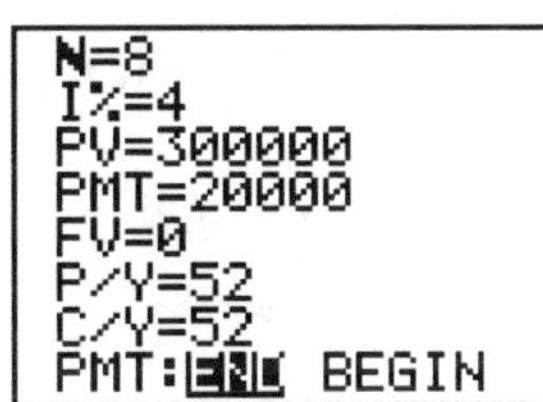

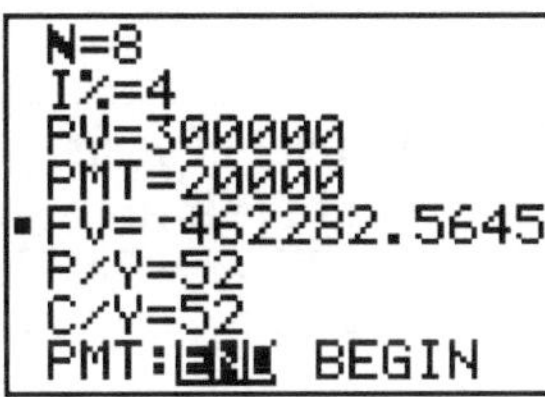

Examples from past IB papers

From November 1999 Paper 1

Tony invested CHF 500 in a bank account at a constant rate of interest. The bank calculates his balance at the end of each year, **rounded to two decimal places**, as shown in the table below.

Year	Value at beginning of year	Value at end of year
1st	CHF 500	CHF 540
2nd	CHF 540	CHF 583.20
3rd	CHF 583.2	CHF 629.86
4th	CHF 629.86	CHF 680.25
5th	CHF 680.25	
6th		

(a) What is the rate and type of interest?

(b) Complete the table for the fifth and sixth year of investment.

Working:

(a) $\frac{540}{500} = 1.08$

$\frac{583.20}{540} = 1.08$

so this is compound interest at 8% per year

(b) $680.25 \times 1.08 = 734.67$

$734.67 \times 1.08 = 793.44$

Answers:

(a) Compound interest, 8% per year

(b) 5th year: CHF 734.67

6 th year: CHF 734.67; CHF 793.44

Now you practise it

From Specimen 2000 Paper 1

The table below gives the monthly repayments for a loan of 1000 Australian dollars (AUD). The interest rates are 18% per annum and 18.5% per annum, respectively.

	Monthly repayments on AUD 1000	
Number of years of the loan	**18% p.a.**	**18.5% p.a.**
1	94.75	95.07
2	51.45	51.72
3	37.15	37.43
4	30.11	30.39
5	26.14	26.26

From the table,

(a) find the monthly repayment on a loan of AUD 1000 at 18.5% per annum taken over 1 year;

(b) calculate the total amount to be repaid on a loan of AUD 8000 taken over 5 years, if the interest rate is 18% per annum.

Working:

Answers:

(a) ....

(b) ....

Additional practice problems

1. From May 2001 Paper 2 *[Maximum mark: 18]*

(i) The following is a currency conversion table:

	FFR	USD	JPY	GBP
French Francs (FFR)	1	p	q	0.101
US Dollars (USD)	6.289	1	111.111	0.631
Japanese Yen (JPY)	0.057	0.009	1	0.006
British Pounds (GBP)	9.901	1.585	166.667	1

For example, from the table 1 USD = 0.631 GBP. Use the table to answer the following questions.

(a) Find the values of p and q. *[2 marks]*

(b) Mireille wants to change money at a bank in London.

(i) How many French Francs (FFR) will she have to change to receive 140 British Pounds (GBP)?

(ii) The bank charges a 2.4% commission on all transactions. If she makes this transaction, how many British Pounds will Mireille actually receive from the bank? *[4 marks]*

(c) Jean invested 5000 FFR in Paris at 8% simple interest per annum. Paul invested 800 GBP in London at 6% simple interest per annum.

(i) How much interest in FFR did Jean earn after 4 years?

(ii) How much interest in US Dollars did Paul earn after 4 years?

(iii) Who had earned more interest after 4 years?

(iv) Explain your reasoning in part (c) (iii). *[7 marks]*

(ii) Takaya invested 1000 JPY at 6.3% simple interest for 15 years. Morimi invested 900 JPY at 6.3% interest compounded annually for 15 years. Who had more money at the end of the 15th year? Justify your answer **clearly**. *[5 marks]*

2. From November 1999 Paper 2 *[Maximum mark: 9]*

Angela needs \$4000 to pay for a car. She was given two options by the car seller.

Option A: *Outright loan*

A loan of \$4000 at a rate of 12% per annum compounded monthly.

(a) Find

(i) the cost of this loan for one year; *[2 marks]*

(ii) the equivalent annual simple interest rate. *[2 marks]*

Option B: *Friendly Credit Terms*

A 25% deposit, followed by 12 equal monthly payments of \$287.50.

(b) (i) How much is to be paid as a deposit under this option? *[1 mark]*

(ii) Find the cost of the loan under *Friendly Credit Terms*. *[2 marks]*

(c) Give a reason why Angela might choose

(i) **Option A**;

(ii) **Option B**. *[2 marks]*

3. From November 2001 Paper 2 *[Maximum mark: 9]*

The table below shows the deposits, in Australian dollars (AUD), made by Vicki in an investment account on the **first** day of each month for the first four months in 1999. The interest rate is 0.75% **per month** compounded monthly. The interest is added to the account at the **end** of each month.

Month	Deposit (AUD)
January	600
February	1300
March	230
April	710

(a) Show that the amount of money in Vicki's account at the **end** of February is AUD 1918.78. *[3 marks]*

(b) Calculate the amount of Australian dollars in Vicki's account at the **end** of April. *[2 marks]*

Vicki makes no withdrawals or deposits after 1st April 1999.

(c) How much money is in Vicki's account at the end of December 1999? *[2 marks]*

From 1st January 2000 the bank applies a new interest rate of 3.5% **per annum** compounded annually.

(d) In how many full years after December 1999 will Vicki's investment first exceed AUD 3300? *[2 marks]*

4. From November 2002 Paper 2 [Maximum mark: 14]

On Vera's 18th birthday she was given an allowance from her parents. She was given the following choices.

Choice A $100 every month of the year.

Choice B A fixed amount of $1100 at the beginning of the year, to be invested at an interest rate of 12% per annum, compounded monthly.

Choice C $75 the first month and an increase of $5 every month thereafter.

Choice D $80 the first month and an increase of 5% every month.

(a) Assuming that Vera does not spend any of her allowance during the year, calculate, for each of the choices, how much money she would have at the end of the year. *[8 marks]*

(b) Which of the choices do you think that Vera should choose? Give a reason for your answer. *[2 marks]*

(c) On her 19th birthday Vera invests $1200 in a bank that pays interest at r% per annum compounded annually. Vera would like to buy a scooter costing $1452 on her 21st birthday. What rate will the bank have to offer her to enable her to buy the scooter? *[4 marks]*

Practice exam 1

Paper 1

LEVEL1DIFFICULTY/I/LOVE/WRITING/IB/CODES

MATHEMATICAL STUDIES

STANDARD LEVEL

PAPER 1

(whenever you feel like taking it!)

1 hour 30 minutes

This is NOT an official IB examination. It is written by the author and has not been approved in any way by the IB.

Candidate session number

0	0							

INSTRUCTIONS TO CANDIDATES

- Write your session number in the boxes above.
- Do not open this examination paper until instructed to do so.
- Answer all the questions in the spaces provided.
- Unless otherwise stated in the question, all numerical answers must be given exactly or correct to three significant figures.

Paper 1 (continued)

1. Donella is studying IB Theatre Arts and wants to determine how brightly to illuminate her stage. She looks in her textbook and runs across this formula:
stage width (in metres) $\times$ stage length (in metres) $\times$ 0.14 = total watts for stage
If her stage is 7.25 metres wide and 11.5 metres long,

(a) calculate the **exact** number of watts needed to illuminate the stage ; *[2 marks]*
(b) calculate the number of watts needed to illuminate the stage, **rounded to the nearest 10 watts** ; *[2 marks]*
(c) write the answer to **part (a)** in the form $a \times 10^k$ where $1 \leq a < 10$ and $k \in \mathbb{Z}$. *[2 marks]*

Working:

Answers:

(a) ....

(b) ....

(c) ....

2. The numbers 3, 6, 9, 12, . . . are in an arithmetic progression.

(a) Write down the common difference. *[1 mark]*
(b) Find the 35th term of the sequence. *[2 marks]*
(c) Find the sum of the first 100 terms of the sequence. *[3 marks]*

Working:

Answers:

(a) ....

(b) ....

(c) ....

3. Chasah decides to take her most recent math studies quiz grades and make a box-and-whisker plot from the data (nobody knows why!). Here are her grades:

54, 36, 97, 24, 10, 45, 72, 74, 91, 62

(a) Write down the median quiz grade. *[1 mark]*
(b) Write down the 25th percentile quiz grade. *[2 marks]*
(c) Given that the 75th percentile quiz grade is 74, draw a box-and-whisker plot of Chasah's quiz grades below. *[3 marks]*

Working:

Answers:

(a) ....

(b) ....

4. A function is defined as follows: $f(x) = x^2 + x - 6$.

(a) Find $f(-3)$. *[1 mark]*
(b) Factorise the expression $x^2 + x - 6$. *[2 marks]*
(c) **Hence or otherwise** write down the coordinates of the points where $f(x) = 0$. *[3 marks]*

Working:

Answers:

(a) ....

(b) ....

(c) ....

5.

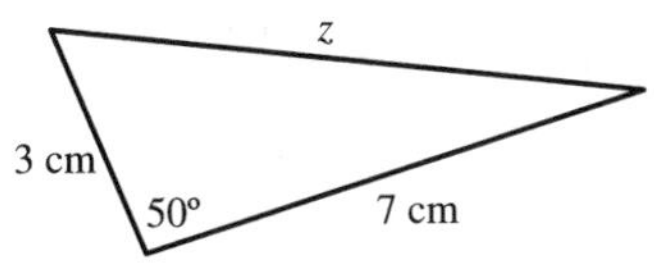

(a) Calculate the length z of the triangle. *[3 marks]*
(b) Calculate the area of the triangle, correct to **2 decimal places.** *[3 marks]*

Working:

Answers:

(a) ....

(b) ....

Paper 1 (continued)

6. Manal puts $3000 in an account earning interest at 11% per annum compounded monthly. She leaves her money in the account for 15 years. How much money does she have in her account if

(a) she leaves her money as it is for 15 years? *[3 marks]*
(b) after ten years, she spends half the money and continues earning **simple** interest at 11% per annum on the remaining half of the money for the remaining 5 years? *[3 marks]*

Working:

Answers:

(a) ....

(b) ....

7. Jed and Caroline are having an argument, as usual. Caroline says to Jed, "If you insult me one more time, I am going to slap you!" Jed replies, "That's ok, Caroline. If I don't insult you, you will be able to slap me anyway." If p = you insult me one more time and q = I am going to slap you,

(a) write Caroline's statement **in symbolic form** ; *[2 marks]*
(b) write **in words** the converse of Caroline's statement ; *[2 marks]*
(c) Explain briefly, according to symbolic logic, why Jed is correct. *[2 marks]*

Working:

Answers:

(a) ....

(b) ....

(c) ....

8. Two functions are defined as follows:

$$f(x) = \frac{3}{x + 1} \text{ and } g(x) = x - 6$$

(a) Write the equations of the vertical and horizontal asymptotes of f. *[3 marks]*
(b) Determine the coordinates of the points where $f(x) = g(x)$. *[3 marks]*

Working:

Answers:

(a) ....

(b) ....

9. Fatema is doing a math studies project on sleep deprivation for IB Diploma students writing their extended essays. Her research question is, "Does the length of the extended essay depend on the number of hours of sleep the night before it is due?" Fatema decides to perform a χ^2 independence test at a 0.05 level of significance and starts by collecting the following data:

		less than 3000 words	more than 3000 words
hours of sleep	$x < 4$	16	10
	$4 \le x \le 6$	14	11
	$x > 6$	8	15

(a) State the null hypothesis for this test. *[1 marks]*

(b) Calculate the expected number of students who get more than 6 hours of sleep and who have essays less than 3000 words. *[2 marks]*

(c) If the χ^2 calculated value is 3.82, explain whether or not the length of the extended essay is dependent on the number of hours of sleep the night before it is due. *[3 marks]*

Working:

Answers:

(a) ....

(b) ....

(c) ....

....

Paper 1 (continued)

10. Francesca, Linda, Hana, Eliza and Sarah decide to travel around Europe after they finish their IB examinations. They start their vacation with 3000 US dollars (USD).

(a) Their first stop is Italy where they spend 800 Euros (EUR) on food and lodging. Given the exchange rate of 1 USD = 0.6848 EUR, calculate how much money they have left after their visit to Italy in USD **rounded to 2 decimal places.** *[2 marks]*

(b) Their second stop is Britain where they find an exchange bureau that offers the following rates for one British pound (GBP):

	London Currency Exchange	
	BUY	SELL
US Dollars (USD)	1.7892	1.7662
Euro (EUR)	1.2379	1.2311
South African Rand (ZAR)	14.891	14.728

Calculate the amount of USD they would need to exchange in order to pay for a 70 GBP night out on the town, **rounded to two decimal places.** *[2 marks]*

(c) At the end of their trip they are left with 400 USD and they use it to purchase a large fake French sculpture. If the value of the sculpture appreciates at a rate of 14% per annum, calculate the current value of the sculpture after 4 years. *[2 marks]*

Working:

Answers:

(a) ....

(b) ....

(c) ....

11. Shade each Venn diagram below according to its respective set notation.

(a)

$(A \cap B)$

[1 mark]

(b)

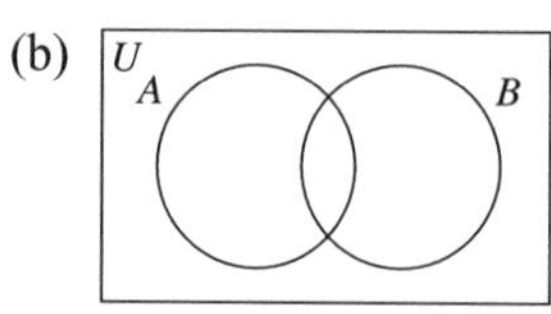

(A')

[1 mark]

(c)

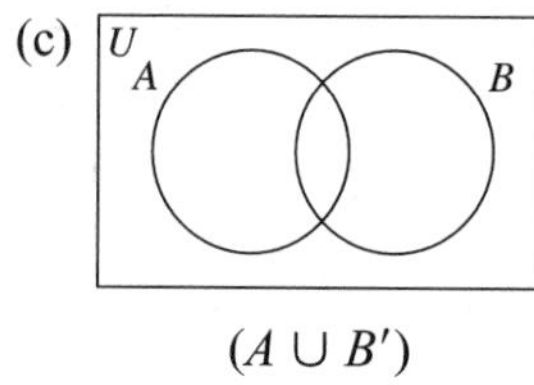

$(A \cup B')$

[2 marks]

(d)

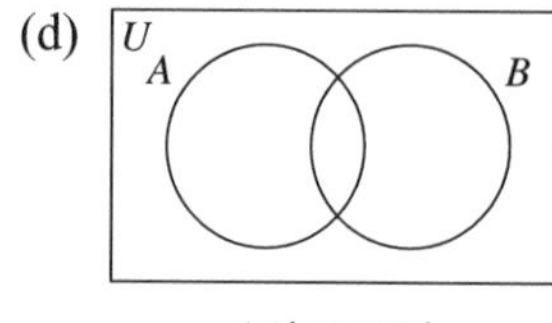

$(A' \cap B)'$

[2 marks]

Working:

12. The graph of the function $f(x) = a \cos bx + c$ is given below.

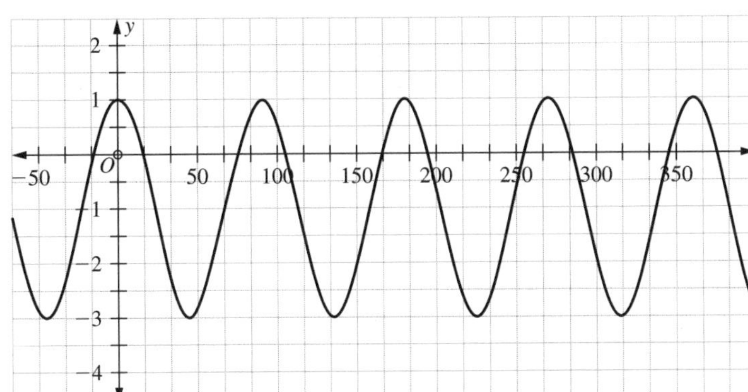

Find the value of

(a) a *[2 marks]*

(b) b *[2 marks]*

(c) c *[2 marks]*

Working:

Answers:

(a) ....

(b) ....

(c) ....

Paper 1 (continued)

13. Given the following system of linear equations

$$2x - y = 7$$
$$3x + 2y = 14$$

find the value of

(a) x *[3 marks]*

(b) y *[3 marks]*

Working:

Answers:

(a) ....

(b) ....

14. Two logical statements are given below:

p = Katie has brown hair. q = Katie is good at mathematics.

(a) Write the logical statement $\neg(p \Leftrightarrow q)$ **in words.** *[2 marks]*

(b) Complete the truth table below.

p	q	$p \Leftrightarrow q$	$\neg(p \Leftrightarrow q)$
T	T	T	
T	F		
F	T		
F	F		F

[2 marks]

(c) State whether or not $\neg(p \Leftrightarrow q)$ is a tautology and explain your reasoning. *[2 marks]*

Working:

Answers:

(a) ....

(c) ....

....

Paper 1 (continued)

15. Sarah and Salome are trying to decide who lives closer to their school. They look at a map of their neighbourhood.

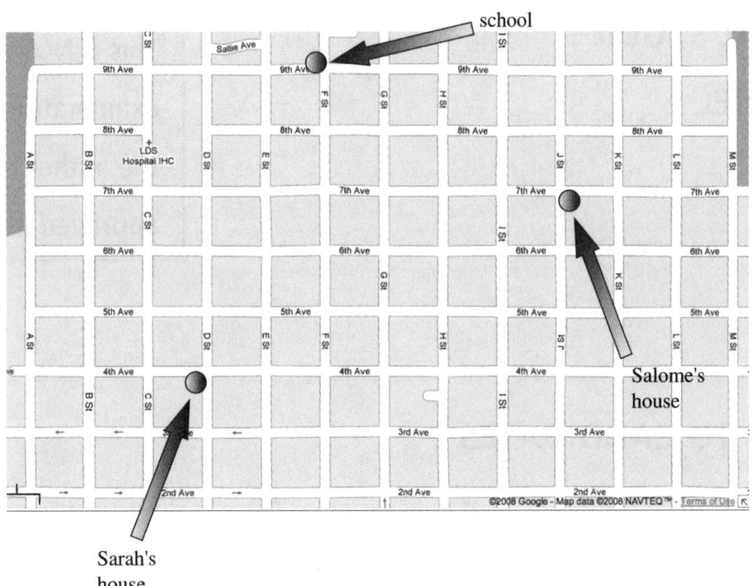

Assuming that the neighbourhood is a perfect coordinate plane where the distance between any two streets or avenues is 150 metres and the coordinates of Salome's house are (1, 2),

(a) write the coordinates of Sarah's house ; *[1 mark]*
(b) calculate the distance from Salome's house to the school, **rounded to the nearest metre** ; *[3 marks]*
(c) show that Salome lives closer to school than Sarah. *[2 marks]*

Working:

Answers:

(a) ....

(b) ....

(c) ....

....

LEVEL1DIFFICULTY/I/LOVE/WRITING/IB/CODES

MATHEMATICAL STUDIES

STANDARD LEVEL

PAPER 2

This is NOT an official IB examination. It is written by the author and has not been approved in any way by the IB.

(whenever you feel like taking it!)

1 hour 30 minutes

INSTRUCTIONS TO CANDIDATES

- Do not open this examination paper until instructed to do so.
- Answer all the questions.
- Unless otherwise stated in the question, all numerical answers must be given exactly or correct to three significant figures.

Paper 2 (continued)

1. *[Maximum mark: 17]*

The slot machines at a casino use three dials. Each dial has 10 facets, on which four symbols (dollars, cherries, trees, and bananas) are unevenly distributed.

Dial	1	2	3
dollars	1 facet	1 facet	1 facet
cherries	3 facets	2 facets	4 facets
trees	3 facets	2 facets	3 facets
bananas	3 facets	5 facets	2 facets

A gambler plays by pulling a handle that spins all three dials. When the dials stop spinning, each dial will show one facet. The gambler wins \$100 000 if all three dials show the dollar. The gambler wins \$10 000 if the dials show all cherries, all trees, or all bananas.

(i) Calculate the probability that Dial 2 shows a dollar. *[1 mark]*
(ii) Given that Dial 3 does not show a dollar, calculate the probability that it shows a cherry. *[2 marks]*
(iii) Calculate the probability that the gambler wins \$100 000. *[3 marks]*
(iv) Calculate the probability that the gambler wins \$10 000. *[4 marks]*
(v) Calculate the probability that the gambler does not win any money. *[3 marks]*

The gambler now pulls the handle two times, each time recording what each dial shows.

(vi) Calculate the probability that the gambler wins \$10 000 on at least one of the two times. *[4 marks]*

2. *[Maximum mark: 13]*

The function, $f(x)$ is defined by $f(x) = x^2 - 2x - 2$.

(a) Find $f'(x)$. *[2 marks]*
(b) Find

(i) the gradient of the tangent to f at the coordinates $(0, -2)$
(ii) the equation of the tangent to f at that point. *[5 marks]*

(c) Use $f'(x)$ to find the coordinates of the vertex of f and state the gradient of the tangent at that point. *[4 marks]*
(d) Write down the intervals when f is

(i) increasing ;
(ii) decreasing. *[2 marks]*

3. *[Maximum mark: 14]*

It is said that certain populations grow so fast that their numbers can be modelled using exponential functions. One such population is known as Harmful Algal Blooms (sometimes called "Red Tides") consisting of microscopic algae such as *Karenia brevis* that multiply very rapidly in water. One red tide was modelled using the following function:

$$A(t) = 3 \cdot (1.4)^t$$

where A is the area of the algal bloom in square metres and t is time in days.

(a) (i) Complete the table below.

t	0	1	2	3	4	5	10	12
A								

(ii) Graph $A(t)$ on separate paper, using 1 cm = 1 day on the horizontal axis and 1 cm = 10 square metres on the vertical axis. *[5 marks]*

(b) Use your graph to determine the area of the algal bloom after 7 days. *[2 marks]*
(c) Use your graph to calculate how long it would take for the algal bloom to have an area of 100 m^2. *[2 marks]*

Paper 2 (continued)

In order to counteract the algal bloom, marine biologists introduce a substance into the ocean that will dissolve the algae. Its area can be modelled using the equation

$$B(t) = 23x + 50$$

(d) Graph $B(x)$ on the same set of axes as $A(x)$. *[3 marks]*

(e) Determine how many days it would take for the area of the counteractive substance to equal the area of the algal bloom. *[2 marks]*

4. *[Maximum mark: 21]*

(a) Lorenzo has a cylindrical block of wood of diameter 6 cm and height 18 cm. He glues the base to a board, but wants to paint the rest of the block. He wants to paint the bottom half of the wood black and the top half white.

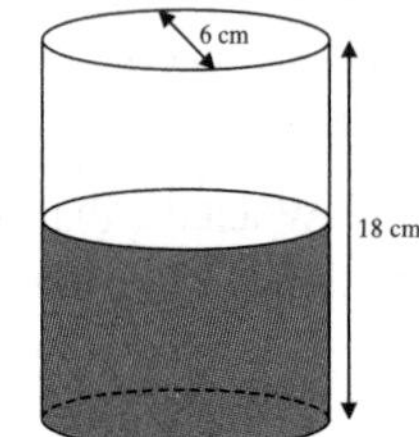

Figure is not drawn to scale.

(i) Calculate the surface area of the block that Lorenzo will paint black. *[3 marks]*

(ii) Calculate the surface area of the block that Lorenzo will paint white. *[3 marks]*

The black paint will cost Lorenzo \$0.03 for every 12 cm^2 painted and the white paint will cost Lorenzo \$0.04 for every 15 cm^2 painted.

(iii) If b represents the area of the black surface and w represents the white surface, write an expression for the cost C of painting the block in terms of b and w. *[2 marks]*

(iv) Calculate the value of C for Lorenzo's block. *[2 marks]*

(v) Determine whether or not Lorenzo could paint his entire block white for less than or equal to \$1.00. *[2 marks]*

(b) Maria has a rectangular block of wood of length 6 cm, width 7 cm, and height 8 cm already painted white. She wants to draw triangles on it from corner to corner.

(i) How many centimetres long is the line from C to F? *[2 marks]*

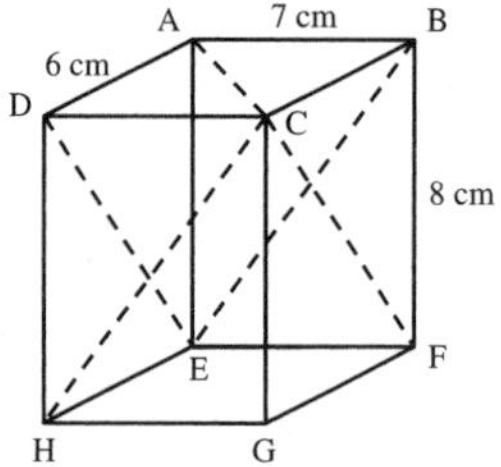

(ii) Find the measure of $\hat{CFG}$. *[2 marks]*

(iii) If the measure of $\hat{FCH}$ is 53°, calculate the measure of $\hat{HFC}$. *[5 marks]*

5. *[Maximum mark: 15]*

The world record time in the 100 metre men's freestyle swimming race has been dropping constantly over the past 100 years. Here are some data from this event:

Year	1935	1954	1961	1972	1976	1985	2000	2008
x	0							
Time t (in sec)	56.6	54.8	53.6	51.47	50.39	48.95	47.84	47.05

(a) Write down the mean and standard deviation for the time for this event. *[2 marks]*

(b) If x represents the number of years since 1935, complete the table for x. *[1 mark]*

Given that the covariance $S_{xt} = -71.129$, find

(c) (i) r, the Pearson product-moment correlation coefficient, for x and t; *[3 marks]*

(ii) the regression equation for t in terms of x; *[4 marks]*

(iii) an estimate for the world record time in 1980, **rounded to four significant figures**. *[4 marks]*

(d) Explain why the regression equation found in part (c) cannot be used to predict a world record time in 2050. *[1 mark]*

Formulae

Presumed knowledge

Area of a parallelogram	$A = (b \times h)$, where b is the base, h is the height
Area of a triangle	$A = \frac{1}{2}(b \times h)$, where b is the base, h is the height
Area of a trapezium	$A = \frac{1}{2}(a+b)h$, where a and b are the parallel sides, h is the height
Area of a circle	$A = \pi r^2$, where r is the radius
Circumference of a circle	$C = 2\pi r$, where r is the radius

Topic 2 – Number and algebra

2.2	*Percentage error*	$\varepsilon = \frac{v_A - v_E}{v_E} \times 100\%$, where v_E is the exact value and v_A is the approximate value of v
2.5	The nth term of an arithmetic sequence	$u_n = u_1 + (n - 1)d$
	The sum of n terms of an arithmetic sequence	$S_n = \frac{n}{2}(2u_1 + (n-1)d) = \frac{n}{2}(u_1 + u_n)$
2.6	The nth term of a geometric sequence	$u_n = u_1 r^{n-1}$
	The sum of n terms of a geometric sequence	$S_n = \frac{u_1(r^n - 1)}{r - 1} = \frac{u_1(1 - r^n)}{1 - r}, r \neq 1$
2.7	Solution of a quadratic equation	$ax^2 + bx + c = 0 \Rightarrow x = \frac{-b \pm \sqrt{b^2 - 4ac}}{2a}, a \neq 0$

Topic 3 – Sets, logic and probability

3.8	Probability of an event A	$P(A) = \frac{n(A)}{n(U)}$
	Complementary events	$P(A') = 1 - P(A)$
3.10	Combined events	$P(A \cup B) = P(A) + P(B) - P(A \cap B)$
	Mutually exclusive events	$P(A \cup B) = P(A) + P(B)$
	Independent events	$P(A \cap B) = P(A)\,P(B)$
	Conditional probability	$P(A \mid B) = \frac{P(A \cap B)}{P(B)}$

Topic 4 – Functions

4.3	Equation of axis of symmetry	$x = -\frac{b}{2a}$

Topic 5 – Geometry and trigonometry

5.1	Distance between two points (x_1,y_1) and (x_2,y_2)	$d = \sqrt{(x_1 - x_2)^2 + (y_1 - y_2)^2}$
	Coordinates of the midpoint of a line segment with endpoints (x_1,y_1) and (x_2,y_2)	$\left(\frac{x_1 + x_2}{2}, \frac{y_1 + y_2}{2}\right)$
5.2	Equation of a straight line	$y = mx + c$; $ax + by + d = 0$
	Gradient formula	$m = \frac{y_2 - y_1}{x_2 - x_1}$
5.4	Sine rule	$\frac{a}{\sin A} = \frac{b}{\sin B} = \frac{c}{\sin C}$
	Cosine rule	$a^2 = b^2 + c^2 - 2bc \cos A$; $\cos A = \frac{b^2 + c^2 - a^2}{2bc}$
	Area of a triangle	Area of a triangle = $\frac{1}{2}ab\sin C$, where a and b are adjacent sides, C is the included angle
5.5	Volume of a pyramid	$V = \frac{1}{3}$(area of base × vertical height)
	Volume of a cuboid	$V = l \times w \times h$, where l is the length, w is the width, h is the height
	Volume of a cylinder	$V = \pi r^2 h$, where r is the radius, h is the height
	Area of the curved surface of a cylinder	$A = 2\pi rh$, where r is the radius, h is the height
	Volume of a sphere	$V = \frac{4}{3}\pi r^3$, where r is the radius
	Surface area of a sphere	$A = 4\pi r^2$, where r is the radius
	Volume of a cone	$V = \frac{1}{3}\pi r^2 h$, where r is the radius, h is the height
	Area of the curved surface of a cone	πrl, where r is the radius, l is the slant height

Topic 6 – Statistics

6.4	Interquartile range	IQR = $q3 - q1$ Outliers are points with value either less than $q1 - 1.5\text{IQR}$ or greater than $q_3 + 1.5\text{IQR}$
6.5	Mean	$\bar{x} = \frac{\sum_{i=1}^{k} f_i x_i}{n}$, where $n = \sum_{i=1}^{k} f_i$
6.6	Standard deviation	$s_n = \sqrt{\frac{\sum_{i=1}^{k} f_i (x_i - \bar{x})^2}{n}}$, where $n = \sum_{i=1}^{k} f_i$
6.7	Pearson's product–moment correlation coefficient	$r = \frac{s_{xy}}{s_x s_y}$, where $s_x = \sqrt{\frac{\sum_{i=1}^{n} (x_i - \bar{x})^2}{n}}$, $s_y = \sqrt{\frac{\sum_{i=1}^{n} (y_i - \bar{y})^2}{n}}$ and s_{xy} *is the covariance*
6.8	Equation of regression line for y on x	$y - y = \frac{s_{xy}}{s_x^2}(x - x)$
6.9	The χ^2 test statistic	$\chi^2_{calc} = \sum \frac{(f_0 - f_e)^2}{f_e}$, where f_0 are the observed frequencies, f_e are the expected frequencies

Topic 7 – Introductory differential calculus

7.1	Derivative of f(x)	$y = f(x) \Rightarrow \frac{\text{d}y}{\text{d}x} = f'(x) \lim_{h \to 0} \left(\frac{f(x+h) - f(x)}{\text{h}} \right)$
7.2	Derivative of ax^n Derivative of a polynomial	$f(x) = ax^n \Rightarrow f'(x) = nax^{n-1}$ $f(x) = ax^n + \text{b}x^{n-1} + \ldots \Rightarrow f'(x) = nax^{n-1} + (n-1)\text{b}x^{n-2} + \ldots$

Topic 8 – Financial mathematics

8.2	Simple interest	$I = \frac{Crn}{100}$, where C = capital, $r\%$ = interest rate, n = number of time periods, I = interest
8.3	Compound interest	$I = C\left(1 + \frac{r}{100}\right)^{\text{n}} - C$, where C = capital, r% = interest rate, n = number of time periods, I = interest which is the special case $k = 1$ of $I = C\left(1 + \frac{r}{100k}\right)^{kn} - C$, where C = capital, $r\%$ = nominal interest rate, n = number of years, k = number of compounding periods in a single year and I = interest

Critical values of the χ^2 distribution (topic 6.9)

$p = P(X \leq c)$

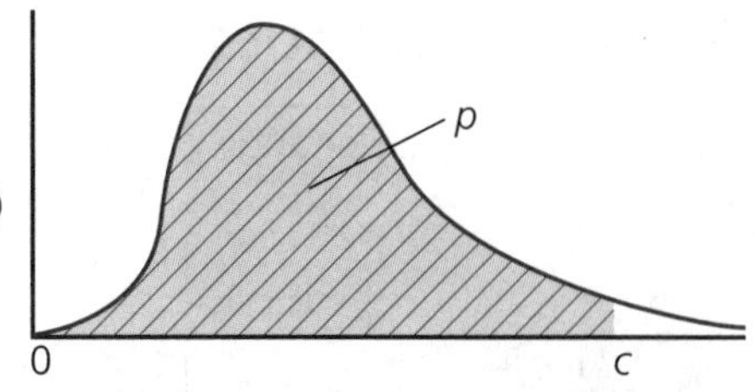

p	0.005	0.01	0.025	0.05	0.1	0.9	0.95	0.975	0.99	0.995
$\nu = 1$	0.00004	0.0002	0.001	0.004	0.016	2.706	3.841	5.024	6.635	7.879
2	0.010	0.020	0.051	0.103	0.211	4.605	5.991	7.378	9.210	10.597
3	0.072	0.115	0.216	0.352	0.584	6.251	7.815	9.348	11.345	12.838
4	0.207	0.297	0.484	0.711	1.064	7.779	9.488	11.143	13.277	14.860
5	0.412	0.554	0.831	1.145	1.610	9.236	11.070	12.833	15.086	16.750
6	0.676	0.872	1.237	1.635	2.204	10.645	12.592	14.449	16.812	18.548
7	0.989	1.239	1.690	2.167	2.833	12.017	14.067	16.013	18.475	20.278
8	1.344	1.646	2.180	2.733	3.490	13.362	15.507	17.535	20.090	21.955
9	1.735	2.088	2.700	3.325	4.168	14.684	16.919	19.023	21.666	23.589
10	2.156	2.558	3.247	3.940	4.865	15.987	18.307	20.483	23.209	25.188
11	2.603	3.053	3.816	4.575	5.578	17.275	19.675	21.920	24.725	26.757
12	3.074	3.571	4.404	5.226	6.304	18.549	21.026	23.337	26.217	28.300
13	3.565	4.107	5.009	5.892	7.042	19.812	22.362	24.736	27.688	29.819
14	4.075	4.660	5.629	6.571	7.790	21.064	23.685	26.119	29.141	31.319
15	4.601	5.229	6.262	7.261	8.547	22.307	24.996	27.488	30.578	32.801
16	5.142	5.812	6.908	7.962	9.312	23.542	26.296	28.845	32.000	34.267
17	5.697	6.408	7.564	8.672	10.085	24.769	27.587	30.191	33.409	35.718
18	6.265	7.015	8.231	9.390	10.865	25.989	28.869	31.526	34.805	37.156
19	6.844	7.633	8.907	10.117	11.651	27.204	30.144	32.852	36.191	38.582
20	7.434	8.260	9.591	10.851	12.443	28.412	31.410	34.170	37.566	39.997
21	8.034	8.897	10.283	11.591	13.240	29.615	32.671	35.479	38.932	41.401
22	8.643	9.542	10.982	12.338	14.041	30.813	33.924	36.781	40.289	42.796
23	9.260	10.196	11.689	13.091	14.848	32.007	35.172	38.076	41.638	44.181
24	9.886	10.856	12.401	13.848	15.659	33.196	36.415	39.364	42.980	45.559
25	10.520	11.524	13.120	14.611	16.473	34.382	37.652	40.646	44.314	46.928
26	11.160	12.198	13.844	15.379	17.292	35.563	38.885	41.923	45.642	48.290
27	11.808	12.879	14.573	16.151	18.114	36.741	40.113	43.195	46.963	49.645
28	12.461	13.565	15.308	16.928	18.939	37.916	41.337	44.461	48.278	50.993
29	13.121	14.256	16.047	17.708	19.768	39.087	42.557	45.722	49.588	52.336
30	13.787	14.953	16.791	18.493	20.599	40.256	43.773	46.979	50.892	53.672
40	20.707	22.164	24.433	26.509	29.051	51.805	55.758	59.342	63.691	66.766
50	27.991	29.707	32.357	34.764	37.689	63.167	67.505	71.420	76.154	79.490
60	35.534	37.485	40.482	43.188	46.459	74.397	79.082	83.298	88.379	91.952
70	43.275	45.442	48.758	51.739	55.329	85.527	90.531	95.023	100.425	104.215
80	51.172	53.540	57.153	60.391	64.278	96.578	101.879	106.629	112.329	116.321
90	59.196	61.754	65.647	69.126	73.291	107.565	113.145	118.136	124.116	128.299
100	67.328	70.065	74.222	77.929	82.358	118.498	124.342	129.561	135.807	140.169

ν = number of degrees of freedom